Contents

Introduction

Of the many diverse and fascinating challenges we face today, the most intense and important is how to understand and shape the new technology revolution, which entails nothing less than a transformation of humankind. We are at the beginning of a revolution that is fundamentally changing the way we live, work, and relate to one another. In its scale, scope and complexity, what I consider to be the fourth industrial revolution is unlike anything humankind has experienced before.

We have yet to grasp fully the speed and breadth of this new revolution. Consider the unlimited possibilities of having billions of people connected by mobile devices, giving rise to unprecedented processing power, storage capabilities and knowledge access. Or think about the staggering confluence of emerging technology breakthroughs, covering wide-ranging fields such as artificial intelligence (AI), robotics, the internet of things (IoT), autonomous vehicles, 3D printing, nanotechnology, biotechnology, materials science, energy storage and quantum computing, to name a few. Many of these innovations are in their infancy, but they are already reaching an inflection point in their development as they build on and amplify each other in a fusion of technologies across the physical, digital and biological worlds.

We are witnessing profound shifts across all industries, marked by the emergence of new business models, the disruption[1] of incumbents and the reshaping of production, consumption, transportation and delivery systems. On the societal front, a paradigm shift is underway in how we work and communicate, as well as how we express, inform and entertain ourselves. Equally, governments and institutions are being reshaped, as are systems of education, healthcare and transportation, among many others. New ways of using technology to change behaviour and our systems of production and consumption also offer the potential for supporting the regeneration and preservation of natural environments, rather than creating hidden costs in the form of externalities.

The changes are historic in terms of their size, speed and scope.

While the profound uncertainty surrounding the development and adoption of emerging technologies means that we do not yet know how the transformations driven by this industrial revolution will unfold, their complexity and interconnectedness across sectors imply that all stakeholders of global society – governments, business, academia, and civil society – have a responsibility to work together to better understand the emerging trends.

Shared understanding is particularly critical if we are to shape a collective future that reflects common objectives and values. We must have a comprehensive and globally shared view of how technology is changing our lives and those of future generations, and how it is reshaping the economic, social, cultural and human context in which we live.

The changes are so profound that, from the perspective of human history, there has never been a time of greater promise or potential peril. My concern, however, is that decision-makers are too often caught in traditional, linear (and non-disruptive) thinking or too absorbed by immediate concerns to think strategically about the forces of disruption and innovation shaping our future.

I am well aware that some academics and professionals consider the developments that I am looking at as simply a part of the third industrial revolution. Three reasons, however, underpin my conviction that a fourth and distinct revolution is underway:

Velocity: Contrary to the previous industrial revolutions, this one is evolving at an exponential rather than linear pace. This is the result of the multifaceted, deeply interconnected world we live in and the fact that new technology begets newer and ever more capable technology.

Breadth and depth: It builds on the digital revolution and combines multiple technologies that are leading to unprecedented paradigm shifts in the economy, business, society, and individually. It is not only changing the "what" and the "how" of doing things but also "who" we are.

Systems Impact: It involves the transformation of entire systems, across (and within) countries, companies, industries and society as a whole.

In writing this book, my intention is to provide a primer on the fourth industrial revolution - what it is, what it will bring, how it will impact us, and what can be done to harness it for the common good. This volume is intended for all those with an interest in our future who are committed to using the opportunities of this revolutionary change to make the world a better place.

I have three main goals:
— to increase awareness of the comprehensiveness and speed of the technological revolution and its multifaceted impact,

– to create a framework for thinking about the technological revolution that outlines the core issues and highlights possible responses, and
– to provide a platform from which to inspire public-private cooperation and partnerships on issues related to the technological revolution.

Above all, this book aims to emphasize the way in which technology and society co-exist. Technology is not an exogenous force over which we have no control. We are not constrained by a binary choice between "accept and live with it" and "reject and live without it". Instead, take dramatic technological change as an invitation to reflect about who we are and how we see the world. The more we think about how to harness the technology revolution, the more we will examine ourselves and the underlying social models that these technologies embody and enable, and the more we will have an opportunity to shape the revolution in a manner that improves the state of the world.

Shaping the fourth industrial revolution to ensure that it is empowering and human-centred, rather than divisive and dehumanizing, is not a task for any single stakeholder or sector or for any one region, industry or culture. The fundamental and global nature of this revolution means it will affect and be influenced by all countries, economies, sectors and people. It is, therefore, critical that we invest attention and energy in multistakeholder cooperation across academic, social, political, national and industry boundaries. These interactions and collaborations are needed to create positive, common and hope-filled narratives, enabling individuals and groups from all parts of the world to participate in, and benefit from, the ongoing transformations.

Much of the information and my own analysis in this book are based on ongoing projects and initiatives of the World Economic Forum and has been developed, discussed and challenged at recent Forum gatherings. Thus, this book also provides a framework for shaping the future activities of the World Economic Forum. I have also drawn from numerous conversations I have had with business, government and civil society leaders, as well as technology pioneers and young people. It is, in that sense, a crowd-sourced book, the product of the collective enlightened wisdom of the Forum's communities.

This book is organized in three chapters. The first is an overview of the fourth industrial revolution. The second presents the main transformative technologies. The third provides a deep dive into the impact of the revolution and some of the policy challenges it poses. I conclude by suggesting practical ideas and solutions on how best to adapt, shape and harness the potential of this great transformation.

1. The Fourth Industrial Revolution

1.1 Historical Context

The word "revolution" denotes abrupt and radical change. Revolutions have occurred throughout history when new technologies and novel ways of perceiving the world trigger a profound change in economic systems and social structures. Given that history is used as a frame of reference, the abruptness of these changes may take years to unfold.

The first profound shift in our way of living – the transition from foraging to farming – happened around 10,000 years ago and was made possible by the domestication of animals. The agrarian revolution combined the efforts of animals with those of humans for the purpose of production, transportation and communication. Little by little, food production improved, spurring population growth and enabling larger human settlements. This eventually led to urbanization and the rise of cities.

The agrarian revolution was followed by a series of industrial revolutions that began in the second half of the 18th century. These marked the transition from muscle power to mechanical power, evolving to where today, with the fourth industrial revolution, enhanced cognitive power is augmenting human production.

The first industrial revolution spanned from about 1760 to around 1840. Triggered by the construction of railroads and the invention of the steam engine, it ushered in mechanical production. The second industrial revolution, which started in the late 19th century and into the early 20th century, made mass production possible, fostered by the advent of electricity and the assembly line. The third industrial revolution began in the 1960s. It is usually called the computer or digital revolution because it was catalysed by the development of semiconductors, mainframe computing (1960s), personal computing (1970s and 80s) and the internet (1990s).

Mindful of the various definitions and academic arguments used to describe the first three industrial revolutions, I believe that today we are at the beginning of a fourth industrial revolution. It began at the turn of this century and builds on the digital revolution. It is characterized by a much more ubiquitous and mobile internet, by smaller and more powerful sensors that have become cheaper, and by artificial intelligence and machine learning.

Digital technologies that have computer hardware, software and networks at their core are not new, but in a break with the third industrial revolution, they are becoming more sophisticated and integrated and are, as a result, transforming societies and the global economy. This is the reason why Massachusetts Institute of Technology (MIT) Professors Erik Brynjolfsson and Andrew McAfee have famously referred to this period as "the second machine age"[2], the title of their 2014 book, stating that the world is at an inflection point where the effect of these digital technologies will manifest with "full force" through automation and and the making of "unprecedented things".

In Germany, there are discussions about "Industry 4.0", a term coined at the Hannover Fair in 2011 to describe how this will revolutionize the organization of global value chains. By enabling "smart factories", the fourth industrial revolution creates a world in which virtual and physical systems of manufacturing globally cooperate with each other in a flexible way. This enables the absolute customization of products and the creation of new operating models.

The fourth industrial revolution, however, is not only about smart and connected machines and systems. Its scope is much wider. Occurring simultaneously are waves of further breakthroughs in areas ranging from gene sequencing to nanotechnology, from renewables to quantum computing. It is the fusion of these technologies and their interaction across the physical, digital and biological domains that make the fourth industrial revolution fundamentally different from previous revolutions.

In this revolution, emerging technologies and broad-based innovation are diffusing much faster and more widely than in previous ones, which continue to unfold in some parts of the world. The second industrial revolution has yet to be fully experienced by 17% of the world as nearly 1.3 billion people still lack access to electricity. This is also true for the third industrial revolution, with more than half of the world's population, 4 billion people, most of whom live in the developing world, lacking internet access. The spindle (the hallmark of the first industrial revolution) took almost 120 years to spread outside of Europe. By contrast, the internet permeated across the globe in less than a decade.

Still valid today is the lesson from the first industrial revolution – that the extent to which society embraces technological innovation is a major determinant of progress. The government and public institutions, as well as the private sector, need to do their part, but it is also essential that citizens see the long-term benefits.

I am convinced that the fourth industrial revolution will be every bit as powerful, impactful and historically important as the previous three. However I have two primary concerns about factors that may limit the potential of the fourth industrial revolution to be effectively and cohesively realized.

First, I feel that the required levels of leadership and understanding of the changes underway, across all sectors, are low when contrasted with the need to rethink our economic, social and political systems to respond to the fourth industrial revolution. As a result, both at the national and global levels, the requisite institutional framework to govern the diffusion of innovation and mitigate the disruption is inadequate at best and, at worst, absent altogether.

Second, the world lacks a consistent, positive and common narrative that outlines the opportunities and challenges of the fourth industrial revolution, a narrative that is essential if we are to empower a diverse set of individuals and communities and avoid a popular backlash against the fundamental changes underway.

1.2 Profound and Systemic Change

The premise of this book is that technology and digitization will revolutionize everything, making the overused and often ill-used adage "this time is different" apt. Simply put, major technological innovations are on the brink of fuelling momentous change throughout the world – inevitably so.

The scale and scope of change explain why disruption and innovation feel so acute today. The speed of innovation in terms of both its development and diffusion is faster than ever. Today's disruptors – Airbnb, Uber, Alibaba and the like – now household names - were relatively unknown just a few years ago. The ubiquitous iPhone was first launched in 2007. Yet there were as many as 2 billion smart phones at the end of 2015. In 2010 Google announced its first fully autonomous car. Such vehicles could soon become a widespread reality on the road.

One could go on. But it is not only speed; returns to scale are equally staggering. Digitization means automation, which in turn means that companies do not incur diminishing returns to scale (or less of them, at least). To give a sense of what this means at the aggregate level, compare Detroit in 1990 (then a major centre of traditional industries) with Silicon Valley in 2014. In 1990, the three biggest companies in Detroit had a combined market capitalization of $36 billion, revenues of $250 billion, and 1.2 million employees. In 2014, the three biggest companies in Silicon Valley had a considerably higher market capitalization ($1.09 trillion), generated roughly the same revenues ($247 billion), but with about 10 times fewer employees (137,000).[3]

The fact that a unit of wealth is created today with much fewer workers compared to 10 or 15 years ago is possible because digital businesses have marginal costs that tend towards zero. Additionally, the reality of the digital age is that many new businesses provide "information goods" with storage, transportation and replication costs that are virtually nil. Some disruptive tech companies seem to require little capital to prosper. Businesses such as Instagram or WhatsApp, for example, did not require much funding to start up, changing the role of capital and scaling business in the context of the fourth industrial revolution. Overall, this shows how returns to scale further encourage scale and influence change across entire systems.

Aside from speed and breadth, the fourth industrial revolution is unique because of the growing harmonization and integration of so many different disciplines and discoveries. Tangible innovations that result from interdependencies among different technologies are no longer science fiction. Today, for example, digital fabrication technologies can interact with the biological world. Some designers and architects are already mixing computational design, additive manufacturing, materials engineering and synthetic biology to pioneer systems that involve the interaction among micro-organisms, our bodies, the products we consume, and even the buildings we inhabit. In doing so, they are making (and even "growing") objects that are continuously mutable and adaptable (hallmarks of the plant and animal kingdoms).[4]

In *The Second Machine Age,* Brynjolfsson and McAfee argue that computers are so dexterous that it is virtually impossible to predict what applications they may be used for in just a few years. Artificial intelligence (AI) is all around us, from self-driving cars and drones to virtual assistants and translation software. This is transforming our lives. AI has made impressive progress, driven by exponential increases in computing power and by the availability of vast amounts of data, from software used to discover new drugs to algorithms that predict our cultural interests. Many of these algorithms learn from the "bread crumb" trails of data that we leave in the digital world. This results in new types of "machine learning" and automated discovery that enables "intelligent" robots and computers to self-programme and find optimal solutions from first principles.

Applications such as Apple's Siri provide a glimpse of the power of one subset of the rapidly advancing AI field – so-called intelligent assistants. Only two years ago, intelligent personal assistants were starting to emerge. Today, voice recognition and artificial intelligence are progressing so quickly that talking to computers will soon become the norm, creating what some technologists call ambient computing, in which robotic personal assistants are constantly available to take notes and respond to user queries. Our devices will become an increasing part of our personal ecosystem, listening to us, anticipating our needs, and helping us when required – even if not asked.

Inequality as a systemic challenge

The fourth industrial revolution will generate great benefits and big challenges in equal measure. A particular concern is exacerbated inequality. The challenges posed by rising inequality are hard to quantify as a great majority of us are consumers and producers, so innovation and disruption will both positively and negatively affect our living standards and welfare.

The consumer seems to be gaining the most. The fourth industrial revolution has made possible new products and services that increase at virtually no cost the efficiency of our personal lives as consumers. Ordering a cab, finding a flight, buying a product, making a payment, listening to music or watching a film – any of these tasks can now be done remotely. The benefits of technology for all of us who consume are incontrovertible. The internet, the smart phone and the thousands of apps are making our lives easier, and – on the whole – more productive. A simple device such as a tablet, which we use for reading, browsing and communicating, possesses the equivalent processing power of 5,000 desktop computers from 30 years ago, while the cost of storing information is approaching zero (storing 1GB costs an average of less than $0.03 a year today, compared to more than $10,000 20 years ago).

The challenges created by the fourth industrial revolution appear to be mostly on the supply side – in the world of work and production. Over the past few years, an overwhelming majority of the most developed countries and also some fast-growing economies such as China have experienced a significant decline in the share of labour as a percentage of GDP. Half of this drop is due to the fall in the relative price of investment goods,[5] itself driven by the progress of innovation (which compels companies to substitute labour for capital).

As a result, the great beneficiaries of the fourth industrial revolution are the providers of intellectual or physical capital – the innovators, the investors, and the shareholders, which explains the rising gap in wealth between those who depend on their labour and those who own capital. It also accounts for the disillusionment among so many workers, convinced that their real income may not increase over their lifetime and that their children may not have a better life than theirs.

Rising inequality and growing concerns about unfairness present such a significant challenge that I will devote a section to this in Chapter Three. The concentration of benefits and value in just a small percentage of people is also exacerbated by the so-called platform effect, in which digitally-driven organizations create networks that match buyers and sellers of a wide variety of products and services and thereby enjoy increasing returns to scale.

The consequence of the platform effect is a concentration of few but powerful platforms which dominate their markets. The benefits are obvious, particularly to consumers: higher value, more convenience and lower costs. Yet so too are the societal risks. To prevent the concentration of value and power in just a few hands, we have to find ways to balance the benefits and risks of digital platforms (including industry platforms) by ensuring openness and opportunities for collaborative innovation.

These are all fundamental changes affecting our economic, social and political systems which are difficult to undo, even if the process of globalization itself were to somehow be reversed. The question for all industries and companies, without exception, is no longer "Am I going to be disrupted?" but "When is disruption coming, what form will it take and how will it affect me and my organization?"

The reality of disruption and the inevitability of the impact it will have on us does not mean that we are powerless in face of it. It is our responsibility to ensure that we establish a set of common values to drive policy choices and to enact the changes that will make the fourth industrial revolution an opportunity for all.

2. Drivers

Countless organizations have produced lists ranking the various technologies that will drive the fourth industrial revolution. The scientific breakthroughs and the new technologies they generate seem limitless, unfolding on so many different fronts and in so many different places. My selection of the key technologies to watch is based on research done by the World Economic Forum and the work of several of the Forum's Global Agenda Councils.

2.1 Megatrends

All new developments and technologies have one key feature in common: they leverage the pervasive power of digitization and information technology. All of the innovations described in this chapter are made possible and are enhanced through digital power. Gene sequencing, for example, could not happen without progress in computing power and data analytics. Similarly, advanced robots would not exist without artificial intelligence, which itself, largely depends on computing power.

To identify the megatrends and convey the broad landscape of technological drivers of the fourth industrial revolution, I have organized the list into three clusters: physical, digital and biological. All three are deeply interrelated and the various technologies benefit from each other based on the discoveries and progress each makes.

2.1.1 Physical

There are four main physical manifestations of the technological megatrends, which are the easiest to see because of their tangible nature:

– autonomous vehicles
– 3D printing
– advanced robotics
– new materials

Autonomous vehicles

The driverless car dominates the news but there are now many other autonomous vehicles including trucks, drones, aircrafts and boats. As technologies such as sensors and artificial intelligence progress, the capabilities of all these autonomous machines improve at a rapid pace. It is only a question of a few years before low-cost, commercially available drones, together with submersibles, are used in different applications.

As drones become capable of sensing and responding to their environment (altering their flight path to avoid collisions), they will be able to do tasks such as checking electric power lines or delivering medical supplies in war zones. In agriculture, the use of drones – combined with data analytics – will enable more precise and efficient use of fertilizer and water, for example.

3D printing

Also called additive manufacturing, 3D printing consists of creating a physical object by printing layer upon layer from a digital 3D drawing or model. This is the opposite of subtractive manufacturing, which is how things have been made until now, with layers being removed from a piece of material until the desired shape is obtained. By contrast, 3D printing starts with loose material and then builds an object into a three-dimensional shape using a digital template.

The technology is being used in a broad range of applications, from large (wind turbines) to small (medical implants). For the moment, it is primarily limited to applications in the automotive, aerospace and medical industries. Unlike mass-produced manufactured goods, 3D-printed products can be easily customized. As current size, cost and speed constraints are progressively overcome, 3D printing will become more pervasive to include integrated electronic components such as circuit boards and even human cells and organs. Researchers are already working on 4D, a process that would create a new generation of self-altering products capable of responding to environmental changes such as heat and humidity. This technology could be used in clothing or footwear, as well as in health-related products such as implants designed to adapt to the human body.

Advanced robotics

Until recently, the use of robots was confined to tightly controlled tasks in specific industries such as automotive. Today, however, robots are increasingly used across all sectors and for a wide range of tasks from precision agriculture to nursing. Rapid progress in robotics will soon make collaboration between humans and machines an everyday reality. Moreover, because of other technological advances, robots are becoming more adaptive and flexible, with their structural and functional design inspired by complex biological structures (an extension of a process called biomimicry, whereby nature's patterns and strategies are imitated).

Advances in sensors are enabling robots to understand and respond better to their environment and to engage in a broader variety of tasks such as household chores. Contrary to the past when they had to be programmed through an autonomous unit, robots can now access information remotely via the cloud and thus connect with a network of other robots. When the next generation of robots emerges, they will likely reflect an increasing emphasis on human-machine collaboration. In Chapter Three, I will explore the ethical and psychological questions raised by human-machine relations.

New materials

With attributes that seemed unimaginable a few years ago, new materials are coming to market. On the whole, they are lighter, stronger, recyclable and adaptive. There are now applications for smart materials that are self-healing or self-cleaning, metals with memory that revert to their original shapes, ceramics and crystals that turn pressure into energy, and so on.

Like many innovations of the fourth industrial revolution, it is hard to know where developments in new materials will lead. Take advanced nanomaterials such as graphene, which is about 200-times stronger than steel, a million-times thinner than a human hair, and an efficient conductor of heat and electricity.[6] When graphene becomes price competitive (gram for gram, it is one of the most expensive materials on earth, with a micrometer-sized flake costing more than $1,000), it could significantly disrupt the manufacturing and infrastructure industries.[7] It could also profoundly affect countries that are heavily reliant on a particular commodity.

Other new materials could play a major role in mitigating the global risks we face. New innovations in thermoset plastics, for example, could make reusable materials that have been considered nearly impossible to recycle but are used in everything from mobile phones and circuit boards to aerospace industry parts. The recent discovery of new classes of recyclable thermosetting polymers called polyhexahydrotriazines (PHTs) is a major step towards the circular economy, which is regenerative by design and works by decoupling growth and resource needs.[8]

2.1.2 Digital

One of the main bridges between the physical and digital applications enabled by the fourth industrial revolution is the internet of things (IoT) – sometimes called the "internet of all things". In its simplest form, it can be described as a relationship between things (products, services, places, etc.) and people that is made possible by connected technologies and various platforms.

Sensors and numerous other means of connecting things in the physical world to virtual networks are proliferating at an astounding pace. Smaller, cheaper and smarter sensors are being installed in homes, clothes and accessories, cities, transport and energy networks, as well as manufacturing processes. Today, there are billions of devices around the world such as smart phones, tablets and computers that are connected to the internet. Their numbers are expected to increase dramatically over the next few years, with estimates ranging from several billions to more than a trillion. This will radically alter the way in which we manage supply chains by enabling us to monitor and optimize assets and activities to a very granular level. In the process, it will have transformative impact across all industries, from manufacturing to infrastructure to healthcare.

Consider remote monitoring – a widespread application of the IoT. Any package, pallet or container can now be equipped with a sensor, transmitter or radio frequency identification (RFID) tag that allows a company to track where it is as it moves through the supply chain – how it is performing, how it is being used, and so on. Similarly, customers can continuously track (practically in real time) the progress of the package or document they are expecting. For companies that are in the business of operating long and complex supply chains, this is transformative. In the near future, similar monitoring systems will also be applied to the movement and tracking of people.

The digital revolution is creating radically new approaches that revolutionize the way in which individuals and institutions engage and collaborate. For example, the blockchain, often described as a "distributed ledger", is a secure protocol where a network of computers collectively verifies a transaction before it can be recorded and approved. The technology that underpins the blockchain creates trust by enabling people who do not know each other (and thus have no underlying basis for trust) to collaborate without having to go through a neutral central authority – i.e. a custodian or central ledger. In essence, the blockchain is a shared, programmable, cryptographically secure and therefore trusted ledger which no single user controls and which can be inspected by everyone.

Bitcoin is so far the best known blockchain application but the technology will soon give rise to countless others. If, at the moment, blockchain technology records financial transactions made with digital currencies such as Bitcoin, it will in the future serve as a registrar for things as different as birth and death certificates, titles of ownership, marriage licenses, educational degrees, insurance claims, medical procedures and votes – essentially any kind of transaction that can be expressed in code. Some countries or institutions are already investigating the blockchain's potential. The government of Honduras, for example, is using the technology to handle land titles while the Isle of Man is testing its use in company registration.

On a broader scale, technology-enabled platforms make possible what is now called the on-demand economy (referred to by some as the sharing economy). These platforms, which are easy to use on a smart phone, convene people, assets and data, creating entirely new ways of consuming goods and services. They lower barriers for businesses and individuals to create wealth, altering personal and professional environments.

The Uber model epitomizes the disruptive power of these technology platforms. These platform businesses are rapidly multiplying to offer new services ranging from laundry to shopping, from chores to parking, from home-stays to sharing long-distance rides. They have one thing in common: by matching supply and demand in a very accessible (low cost) way, by providing consumers with diverse goods, and by allowing both parties to interact and give feedback, these platforms therefore seed trust. This enables the effective use of under-utilized assets – namely those belonging to people who had previously never thought of themselves as suppliers (i.e. of a seat in their car, a spare bedroom in their home, a commercial link between a retailer and manufacturer, or the time and skill to provide a service like delivery, home repair or administrative tasks).

The on-demand economy raises the fundamental question: What is worth owning – the platform or the underlying asset? As media strategist Tom Goodwin wrote in a TechCrunch article in March 2015: "Uber, the world's largest taxi company, owns no vehicles. Facebook, the world's most popular media owner, creates no content. Alibaba, the most valuable retailer, has no inventory. And Airbnb, the world's largest accommodation provider, owns no real estate."[9]

Digital platforms have dramatically reduced the transaction and friction costs incurred when individuals or organizations share the use of an asset or provide a service. Each transaction can now be divided into very fine increments, with economic gains for all parties involved. In addition, when using digital platforms, the marginal cost of producing each additional product, good or service tends towards zero. This has dramatic implications for business and society that I will explore in Chapter Three.

2.1.3 Biological

Innovations in the biological realm – and genetics in particular – are nothing less than breath-taking. In recent years, considerable progress has been achieved in reducing the cost and increasing the ease of genetic sequencing, and lately, in activating or editing genes. It took more than 10 years, at a cost of $2.7 billion, to complete the Human Genome Project. Today, a genome can be sequenced in a few hours and for less than a thousand dollars.[10] With advances in computing power, scientists no longer go by trial and error; rather, they test the way in which specific genetic variations generate particular traits and diseases.

Synthetic biology is the next step. It will provide us with the ability to customize organisms by writing DNA. Setting aside the profound ethical issues this raises, these advances will not only have a profound and immediate impact on medicine but also on agriculture and the production of biofuels.

Many of our intractable health challenges, from heart disease to cancer, have a genetic component. Because of this, the ability to determine our individual genetic make-up in an efficient and cost-effective manner (through sequencing machines used in routine diagnostics) will revolutionize personalized and effective healthcare. Informed by a tumour's genetic make-up, doctors will be able to make decisions about a patient's cancer treatment.

While our understanding of the links between genetic markers and disease is still poor, increasing amounts of data will make precision medicine possible, enabling the development of highly targeted therapies to improve treatment outcomes. Already, IBM's Watson supercomputer system can help recommend, in just a few minutes, personalized treatments for cancer patients by comparing the histories of disease and treatment, scans and genetic data against the (almost) complete universe of up-to-date medical knowledge.[11]

The ability to edit biology can be applied to practically any cell type, enabling the creation of genetically modified plants or animals, as well as modifying the cells of adult organisms including humans. This differs from genetic engineering practiced in the 1980s in that it is much more precise, efficient and easier to use than previous methods. In fact, the science is progressing so fast that the limitations are now less technical than they are legal, regulatory and ethical. The list of potential applications is virtually endless – ranging from the ability to modify animals so that they can be raised on a diet that is more economical or better suited to local conditions, to creating food crops that are capable of withstanding extreme temperatures or drought.

As research into genetic engineering progresses (for example, the development of the CRISPR/Cas9 method in gene editing and therapy), the constraints of effective delivery and specificity will be overcome, leaving us with one immediate and most challenging question, particularly from an ethical viewpoint: How will genetic editing revolutionize medical research and medical treatment? In principle, both plants and animals could potentially be engineered to produce pharmaceuticals and other forms of treatment. The day when cows are engineered to produce in its milk a blood-clotting element, which haemophiliacs lack, is not far off. Researchers have already started to engineer the genomes of pigs with the goal of growing organs suitable for human transplantation (a process called xenotransplantation, which could not be envisaged until now because of the risk of immune rejection by the human body and of disease transmission from animals to humans).

In line with the point made earlier about how different technologies fuse and enrich each other, 3D manufacturing will be combined with gene editing to produce living tissues for the purpose of tissue repair and regeneration – a process called bioprinting. This has already been used to generate skin, bone, heart and vascular tissue. Eventually, printed liver-cell layers will be used to create transplant organs.

We are developing new ways to embed and employ devices that monitor our activity levels and blood chemistry, and how all of this links to well-being, mental health and productivity at home and at work. We are also learning far more about how the human brain functions and we are seeing exciting developments in the field of neurotechnology. This is underscored by the fact that – over the past few years - two of the most funded research programs in the world are in brain sciences.

It is in the biological domain where I see the greatest challenges for the development of both social norms and appropriate regulation. We are confronted with new questions around what it means to be human, what data and information about our bodies and health can or should be shared with others, and what rights and responsibilities we have when it comes to changing the very genetic code of future generations.

To return to the issue of genetic editing, that it is now far easier to manipulate with precision the human genome within viable embryos means that we are likely to see the advent of designer babies in the future who possess particular traits or who are resistant to a specific disease. Needless to say, discussions about the opportunities and challenges of these capabilities are underway. Notably, in December 2015, the National Academy of Sciences and National Academy of Medicine of the US, the Chinese Academy of Sciences and the Royal Society of the UK convened an International Summit on Human Gene Editing. Despite such deliberations, we are not yet prepared to confront the realities and consequences of the latest genetic techniques even though they are coming. The social, medical, ethical and psychological challenges that they pose are considerable and need to be resolved, or at the very least, properly addressed.

The dynamics of discovery

Innovation is a complex, social process, and not one we should take for granted. Therefore, even though this section has highlighted a wide array of technological advances with the power to change the world, it is important that we pay attention to how we can ensure such advances continue to be made and directed towards the best possible outcomes.

Academic institutions are often regarded as one of the foremost places to pursue forward-thinking ideas. New evidence, however, indicates that the career incentives and funding conditions in universities today favour incremental, conservative research over bold and innovative programmes.[12]

One antidote to research conservatism in academia is to encourage more commercial forms of research. This too, however, has its challenges. In 2015, Uber Technologies Inc. hired 40 researchers and scientists in robotics from Carnegie Mellon University, a significant proportion of the human capital of a lab, impacting its research capabilities and putting stress on the university's contracts with the U.S. Department of Defence and other organizations.[13]

To foster both ground-breaking fundamental research and innovative technical adaptations across academia and business alike, governments should allocate more aggressive funding for ambitious research programmes. Equally, public-private research collaborations should increasingly be structured towards building knowledge and human capital to the benefit of all.

2.2 Tipping Points

When these megatrends are discussed in general terms, they seem rather abstract. They are, however, giving rise to very practical applications and developments.

A World Economic Forum report published in September 2015 identified 21 tipping points – moments when specific technological shifts hit mainstream society – that will shape our future digital and hyper-connected world.[14] They are all expected to occur in the next 10 years and therefore vividly capture the deep shifts triggered by the fourth industrial revolution. The tipping points were identified through a survey conducted by the World Economic Forum's Global Agenda Council on the Future of Software and Society, in which over 800 executives and experts from the information and communications technology sector participated.

Table 1 presents the percentage of respondents who expect that the specific tipping point will have occurred by 2025.[15] In the Appendix, each tipping point and its positive and negative impacts are presented in more detail. Two tipping points that were not part of the original survey – designer beings and neurotechnologies – are also included but do not appear on Table 1.

These tipping points provide important context as they signal the substantive changes that lie ahead - amplified by their systemic nature - and how best to prepare and respond. As I explore in the next chapter, navigating this transition begins with awareness of the shifts that are going on, as well as those to come, and their impact on all levels of global society.

Table 1: Tipping points expected to occur by 2025

	%
10% of people wearing clothes connected to the internet	91.2
90% of people having unlimited and free (advertising-supported) storage	91.0
1 trillion sensors connected to the internet	89.2
The first robotic pharmacist in the US	86.5
10% of reading glasses connected to the internet	85.5
80% of people with a digital presence on the internet	84.4
The first 3D-printed car in production	84.1
The first government to replace its census with big-data sources	82.9
The first implantable mobile phone available commercially	81.7
5% of consumer products printed in 3D	81.1
90% of the population using smartphones	80.7
90% of the population with regular access to the internet	78.8
Driverless cars equalling 10% of all cars on US roads	78.2
The first transplant of a 3D-printed liver	76.4
30% of corporate audits performed by AI	75.4
Tax collected for the first time by a government via a blockchain	73.1
Over 50% of internet traffic to homes for appliances and devices	69.9
Globally more trips/journeys via car sharing than in private cars	67.2
The first city with more than 50,000 people and no traffic lights	63.7
10% of global gross domestic product stored on blockchain technology	57.9
The first AI machine on a corporate board of directors	45.2

Source: *Deep Shift – Technology Tipping Points and Societal Impact,* Global Agenda Council on the Future of Software and Society, World Economic Forum, September 2015.

3. Impact

The scale and breadth of the unfolding technological revolution will usher in economic, social and cultural changes of such phenomenal proportions that they are almost impossible to envisage. Nevertheless, this chapter describes and analyses the potential impact of the fourth industrial revolution on the economy, business, governments and countries, society and individuals.

In all these areas, one of the biggest impacts will likely result from a single force: empowerment – how governments relate to their citizens; how enterprises relate to their employees, shareholders and customers; or how superpowers relate to smaller countries. The disruption that the fourth industrial revolution will have on existing political, economic and social models will therefore require that empowered actors recognize that they are part of a distributed power system that requires more collaborative forms of interaction to succeed.

3.1 Economy

The fourth industrial revolution will have a monumental impact on the global economy, so vast and multifaceted that it makes it hard to disentangle one particular effect from the next. Indeed, all the big macro variables one can think of – GDP, investment, consumption, employment, trade, inflation and so on – will be affected. I have decided to focus only on the two most critical dimensions: growth (in large part through the lens of its long-term determinant, productivity) and employment.

3.1.1 Growth

The impact that the fourth industrial revolution will have on economic growth is an issue that divides economists. On one side, the techno-pessimists argue that the critical contributions of the digital revolution have already been made and that their impact on productivity is almost over. In the opposite camp, techno-optimists claim that technology and innovation are at an inflection point and will soon unleash a surge in productivity and higher economic growth.

While I acknowledge aspects of both sides of the argument, I remain a pragmatic optimist. I am well aware of the potential deflationary impact of technology (even when defined as "good deflation") and how some of its distributional effects can favour capital over labour and also squeeze wages (and therefore consumption). I also see how the fourth industrial revolution enables many people to consume more at a lower price and in a way that often makes consumption more sustainable and therefore responsible.

It is important to contextualize the potential impacts of the fourth industrial revolution on growth with reference to recent economic trends and other factors that contribute to growth. In the few years before the economic and financial crisis that began in 2008, the global economy was growing by about 5% a year. If this rate had continued, it would have allowed global GDP to double every 14-15 years, with billions of people lifted out of poverty.

In the immediate aftermath of the Great Recession, the expectation that the global economy would return to its previous high-growth pattern was widespread. But this has not happened. The global economy seems to be stuck at a growth rate lower than the post-war average – about 3-3.5% a year.

Some economists have raised the possibility of a "centennial slump" and talk about "secular stagnation", a term coined during the Great Depression by Alvin Hansen, and recently brought back in vogue by economists Larry Summers and Paul Krugman. "Secular stagnation" describes a situation of persistent shortfalls of demand, which cannot be overcome even with near-zero interest rates. Although this idea is disputed among academics, it has momentous implications. If true, it suggests that global GDP growth could decline even further. We can imagine an extreme scenario in which annual global GDP growth falls to 2%, which would mean that it would take 36 years for global GDP to double.

There are many explanations for slower global growth today, ranging from capital misallocation to over indebtedness to shifting demographics and so on. I will address two of them, ageing and productivity, as both are particularly interwoven with technological progress.

Ageing

The world's population is forecast to expand from 7.2 billion today to 8 billion by 2030 and 9 billion by 2050. This should lead to an increase in aggregate demand. But there is another powerful demographic trend: ageing. The conventional wisdom is that ageing primarily affects rich countries in the West. This is not the case, however. Birth rates are falling below replacement levels in many regions of the world – not only in Europe, where the decline began, but also in most of South America and the Caribbean, much of Asia including China and southern India, and even some countries in the Middle East and North Africa such as Lebanon, Morocco and Iran.

Ageing is an economic challenge because unless retirement ages are drastically increased so that older members of society can continue to contribute to the workforce (an economic imperative that has many economic benefits), the working-age population falls at the same time as the percentage of dependent elders increases. As the population ages and there are fewer young adults, purchases of big-ticket items such as homes, furniture, cars and appliances decrease. In addition, fewer people are likely to take entrepreneurial risks because ageing workers tend to preserve the assets they need to retire comfortably rather than set up new businesses. This is somewhat balanced by people retiring and drawing down their accumulated savings, which in the aggregate lowers savings and investment rates.

These habits and patterns may change of course, as ageing societies adapt, but the general trend is that an ageing world is destined to grow more slowly unless the technology revolution triggers major growth in productivity, defined simply as the ability to work smarter rather than harder.

The fourth industrial revolution provides us with the ability to live longer, healthier and more active lives. As we live in a society where more than a quarter of the children born today in advanced economies are expected to live to 100, we will have to rethink issues such the working age population, retirement and individual life-planning.[16] The difficulty that many countries are showing in attempting to discuss these issues is just a further sign of how we are not prepared to adequately and proactively recognize the forces of change.

Productivity

Over the past decade, productivity around the world (whether measured as labour productivity or total-factor productivity (TFP)) has remained sluggish, despite the exponential growth in technological progress and investments in innovation.[17] This most recent incarnation of the productivity paradox – the perceived failure of technological innovation to result in higher levels of productivity – is one of today's great economic enigmas that predates the onset of the Great Recession, and for which there is no satisfactory explanation.

Consider the US, where labour productivity grew on average 2.8 percent between 1947 and 1983, and 2.6 percent between 2000 and 2007, compared with 1.3 percent between 2007 and 2014.[18] Much of this drop is due to lower levels of TFP, the measure most commonly associated with the contribution to efficiency stemming from technology and innovation. The US Bureau of Labour Statistics indicates that TFP growth between 2007 and 2014 was only 0.5%, a significant drop when compared to the 1.4% annual growth in the period 1995 to 2007.[19] This drop in measured productivity is particularly concerning given that it has occurred as the 50 largest US companies have amassed cash assets of more than $1 trillion, despite real interest rates hovering around zero for almost five years.[20]

Productivity is the most important determinant of long-term growth and rising living standards so its absence, if maintained throughout the fourth industrial revolution, means that we will have less of each. Yet how can we reconcile the data indicating declining productivity with the expectations of higher productivity that tend to be associated with the exponential progress of technology and innovation?

One primary argument focuses on the challenge of measuring inputs and outputs and hence discerning productivity. Innovative goods and services created in the fourth industrial revolution are of significantly higher functionality and quality, yet are delivered in markets that are fundamentally different from those which we are traditionally used to measuring. Many new goods and services are "non-rival", have zero marginal costs and/or harness highly-competitive markets via digital platforms, all of which result in lower prices. Under these conditions, our traditional statistics may well fail to capture real increases in value as consumer surplus is not yet reflected in overall sales or higher profits.

Hal Varian, Google's chief economist, points to various examples such as the increased efficiency of hailing a taxi through a mobile app or renting a car through the power of the on-demand economy. There are many other similar services whose use tends to increase efficiency and hence productivity. Yet because they are essentially free, they therefore provide uncounted value at home and at work. This creates a discrepancy between the value delivered via a given service versus growth as measured in national statistics. It also suggests that we are actually producing and consuming more efficiently than our economic indicators suggest.[21]

Another argument is that, while the productivity gains from the third industrial revolution may well be waning, the world has yet to experience the productivity explosion created by the wave of new technologies being produced at the heart of the fourth industrial revolution.

Indeed, as a pragmatic optimist, I feel strongly that we are only just beginning to feel the positive impact on the world that the fourth industrial revolution can have. My optimism stems from three main sources.

First, the fourth industrial revolution offers the opportunity to integrate the unmet needs of two billion people into the global economy, driving additional demands for existing products and services by empowering and connecting individuals and communities all over the world to one another.

Second, the fourth industrial revolution will greatly increase our ability to address negative externalities and, in the process, to boost potential economic growth. Take carbon emissions, a major negative externality, as an example. Until recently, green investing was only attractive when heavily subsidized by governments. This is less and less the case. Rapid technological advances in renewable energy, fuel efficiency and energy storage not only make investments in these fields increasingly profitable, boosting GDP growth, but they also contribute to mitigating climate change, one of the major global challenges of our time.

Third, as I discuss in the next section, businesses, governments and civil society leaders with whom I interact all tell me that they are struggling to transform their organizations to realize fully the efficiencies that digital capabilities deliver. We are still at the beginning of the fourth industrial revolution, and it will require entirely new economic and organizational structures to grasp its full value.

Indeed, my view is that the competitiveness rules of the fourth industrial revolution economy are different from previous periods. To remain competitive, both companies and countries must be at the frontier of innovation in all its forms, which means that strategies which primarily focus on reducing costs will be less effective than those which are based on offering products and services in more innovative ways. As we see today, established companies are being put under extreme pressure by emerging disruptors and innovators from other industries and countries. The same could be said for countries that do not recognize the need to focus on building their innovation ecosystems accordingly.

To sum up, I believe that the combination of structural factors (over-indebtedness and ageing societies) and systemic ones (the introduction of the platform and on-demand economies, the increasing relevance of decreasing marginal costs, etc.) will force us to rewrite our economic textbooks. The fourth industrial revolution has the potential both to increase economic growth and to alleviate some of the major global challenges we collectively face. We need, however, to also recognize and manage the negative impacts it can have, particularly with regard to inequality, employment and labour markets.

3.1.2 Employment

Despite the potential positive impact of technology on economic growth, it is nonetheless essential to address its possible negative impact, at least in the short term, on the labour market. Fears about the impact of technology on jobs are not new. In 1931, the economist John Maynard Keynes famously warned about widespread technological unemployment "due to our discovery of means of economising the use of labour outrunning the pace at which we can find new uses for labour".[22] This proved to be wrong but what if this time it were true? Over the past few years, the debate has been reignited by evidence of computers substituting for a number of jobs, most notably bookkeepers, cashiers and telephone operators.

The reasons why the new technology revolution will provoke more upheaval than the previous industrial revolutions are those already mentioned in the introduction: speed (everything is happening at a much faster pace than ever before), breadth and depth (so many radical changes are occurring simultaneously), and the complete transformation of entire systems.

In light of these driving factors, there is one certainty: New technologies will dramatically change the nature of work across all industries and occupations. The fundamental uncertainty has to do with the extent to which automation will substitute for labour. How long will this take and how far will it go?

To get a grasp on this, we have to understand the two competing effects that technology exercises on employment. First, there is a destruction effect as technology-fuelled disruption and automation substitute capital for labour, forcing workers to become unemployed or to reallocate their skills elsewhere. Second, this destruction effect is accompanied by a capitalization effect in which the demand for new goods and services increases and leads to the creation of new occupations, businesses and even industries.

As human beings, we have an amazing ability for adaptation and ingenuity. But the key here is the timing and extent to which the capitalization effect supersedes the destruction effect, and how quickly the substitution will take.

There are roughly two opposing camps when it comes to the impact of emerging technologies on the labour market: those who believe in a happy ending – in which workers displaced by technology will find new jobs, and where technology will unleash a new era of prosperity; and those who believe it will lead to a progressive social and political Armageddon by creating technological unemployment on a massive scale. History shows that the outcome is likely to be somewhere in the middle. The question is: What should we do to foster more positive outcomes and help those caught in the transition?

It has always been the case that technological innovation destroys some jobs, which it replaces in turn with new ones in a different activity and possibly in another place. Take agriculture as an example. In the US, people working on the land consisted of 90% of the workforce at the beginning of the 19th century, but today, this accounts for less than 2%. This dramatic downsizing took place relatively smoothly, with minimal social disruption or endemic unemployment.

The app economy provides an example of a new job ecosystem. It only began in 2008 when Steve Jobs, the founder of Apple, let outside developers create applications for the iPhone. By mid-2015, the global app economy was expected to generate over $100 billion in revenues, surpassing the film industry, which has been in existence for over a century.

The techno-optimists ask: If we extrapolate from the past, why should it be different this time? They acknowledge that technology can be disruptive but claim that it always ends up improving productivity and increasing wealth, leading in turn to greater demand for goods and services and new types of jobs to satisfy it. The substance of the argument goes as follows: Human needs and desires are infinite so the process of supplying them should also be infinite. Barring the normal recessions and occasional depressions, there will always be work for everybody.

What evidence supports this and what does it tell us about what lies ahead? The early signs point to a wave of labour-substitutive innovation across multiple industries and job categories which will likely happen in the coming decades.

Labour substitution

Many different categories of work, particularly those that involve mechanically repetitive and precise manual labour, have already been automated. Many others will follow, as computing power continues to grow exponentially. Sooner than most anticipate, the work of professions as different as lawyers, financial analysts, doctors, journalists, accountants, insurance underwriters or librarians may be partly or completely automated.

So far, the evidence is this: The fourth industrial revolution seems to be creating fewer jobs in new industries than previous revolutions. According to an estimate from the Oxford Martin Programme on Technology and Employment, only 0.5% of the US workforce is employed in industries that did not exist at the turn of the century, a far lower percentage than the approximately 8% of new jobs created in new industries during the 1980s and the 4.5% of new jobs created during the 1990s. This is corroborated by a recent US Economic Census, which sheds some interesting light on the relationship between technology and unemployment. It shows that innovations in information and other disruptive technologies tend to raise productivity by replacing existing workers, rather than creating new products needing more labour to produce them.

Two researchers from the Oxford Martin School, economist Carl Benedikt Frey and machine learning expert Michael Osborne, have quantified the potential effect of technological innovation on unemployment by ranking 702 different professions according to their probability of being automated, from the least susceptible to the risk of automation ("0" corresponding to no risk at all) to those that are the most susceptible to the risk ("1" corresponding to a certain risk of the job being replaced by a computer of some sort).[23] In Table 2 below, I highlight certain professions that are most likely to be automated, and those least likely.

This research concludes that about 47% of total employment in the US is at risk, perhaps over the next decade or two, characterized by a much broader scope of job destruction at a much faster pace than labour market shifts experienced in previous industrial revolutions. In addition, the trend is towards greater polarization in the labour market. Employment will grow in high-income cognitive and creative jobs and low-income manual occupations, but it will greatly diminish for middle-income routine and repetitive jobs.

Table 2: Examples of professions most and least prone to automation

Most Prone to Automation

Probability	Occupation
0.99	Telemarketers
0.99	Tax preparers
0.98	Insurance Appraisers, Auto Damage
0.98	Umpires, Referees, and Other Sports Officials
0.98	Legal Secretaries
0.97	Hosts and Hostesses, Restaurant, Lounge, and Coffee Shop
0.97	Real Estate Brokers
0.97	Farm Labour Contractors
0.96	Secretaries and Administrative Assistants, Except Legal, Medical & Executive
0.94	Couriers and Messengers

Least Prone to Automation

Probability	Occupation
0.0031	Mental Health and Substance Abuse Social Workers
0.0040	Choreographers
0.0042	Physicians and Surgeons
0.0043	Psychologists
0.0055	Human Resources Managers
0.0065	Computer Systems Analysts
0.0077	Anthropologists and Archeologists
0.0100	Marine Engineers and Naval Architects
0.0130	Sales Managers
0.0150	Chief Executives

Source: Carl Benedikt Frey and Michael Osborne, University of Oxford, 2013

It is interesting to note that it is not only the increasing abilities of algorithms, robots and other forms of non-human assets that are driving this substitution. Michael Osborne observes that a critical enabling factor for automation is the fact that companies have worked hard to define better and simplify jobs in recent years as part of their efforts to outsource, off-shore and allow them to be performed as "digital work" (such as via Amazon's Mechanical Turk, or MTurk, service, a crowdsourcing internet marketplace). This job simplification means that algorithms are better able to replace humans. Discrete, well-defined tasks lead to better monitoring and more high-quality data around the task, thereby creating a better base from which algorithms can be designed to do the work.

In thinking about the automation and the phenomenon of substitution, we should resist the temptation to engage in polarized thinking about the impact of technology on employment and the future of work. As Frey and Osborne's work shows, it is almost inevitable that the fourth industrial revolution will have a major impact on labour markets and workplaces around the world. But this does not mean that we face a man-versus-machine dilemma. In fact, in the vast majority of cases, the fusion of digital, physical and biological technologies driving the current changes will serve to enhance human labour and cognition, meaning that leaders need to prepare workforces and develop education models to work with, and alongside, increasingly capable, connected and intelligent machines.

Impact on skills

In the foreseeable future, low-risk jobs in terms of automation will be those that require social and creative skills; in particular, decision-making under uncertainty and the development of novel ideas.

This, however, may not last. Consider one of the most creative professions – writing – and the advent of automated narrative generation. Sophisticated algorithms can create narratives in any style appropriate to a particular audience. The content is so human-sounding that a recent quiz by The New York Times showed that when reading two similar pieces, it is impossible to tell which one has been written by a human writer and which one is the product of a robot. The technology is progressing so fast that Kristian Hammond, co-founder of Narrative Science, a company specializing in automated narrative generation, forecasts that by the mid-2020s, 90% of news could be generated by an algorithm, most of it without any kind of human intervention (apart from the design of the algorithm, of course).[24]

In such a rapidly evolving working environment, the ability to anticipate future employment trends and needs in terms of the knowledge and skills required to adapt becomes even more critical for all stakeholders. These trends vary by industry and geography, and so it is important to understand the industry and country-specific outcomes of the fourth industrial revolution.

In the Forum's *Future of Jobs Report*, we asked the chief human resources officers of today's largest employers in 10 industries and 15 economies to imagine the impact on employment, jobs and skills up to the year 2020. As Figure 1 shows, survey respondents believe that complex problem solving, social and systems skills will be far more in demand in 2020 when compared to physical abilities or content skills. The report finds that the next five years are a critical period of transition: the overall employment outlook is flat but there is significant job churn within industries and skill churn within most occupations. While wages and work-life balance are expected to improve slightly for most occupations, job security is expected to worsen in half of the industries surveyed. It is also clear that women and men will be affected differently, potentially exacerbating gender inequality (see Box A: Gender Gaps and the Fourth Industrial Revolution).

Figure 1: Skills Demand in 2020

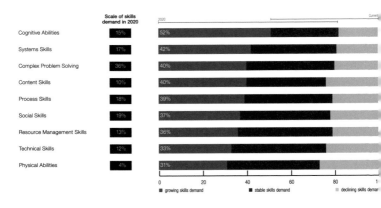

Source: Future of Jobs Report, World Economic Forum

Box A: Gender Gaps and the Fourth Industrial Revolution

The 10th edition of the World Economic Forum's *Global Gender Gap Report 2015* revealed two worrying trends. First, at the current pace of progress, it will take another 118 years before economic gender parity is achieved around the world. Second, progress towards parity is remarkably slow, and possibly stalling.

In light of this, it is critical to consider the impact of the fourth industrial revolution on the gender gap. How will the accelerating pace of change in technologies that span the physical, digital and biological worlds affect the role that women are able to play in the economy, politics and society?

An important question to consider is whether female-dominated or male-dominated professions are more susceptible to automation. The Forum's *Future of Jobs* report indicates that significant job losses are likely to span both types. While there has tended to be more unemployment due to automation in sectors in which men dominate such as manufacturing, construction and installation, the increasing capabilities of artificial intelligence and the ability to digitize tasks in service industries indicate that a wide range of jobs are at risk, from positions at call centres in emerging markets (the source of livelihoods for large numbers of young female workers who are the first in their families to work) to retail and administrative roles in developed economies (a key employer for lower-middle class women).

Losing a job has negative effects in many circumstances, but the cumulative effect of significant losses across whole job categories that have traditionally given women access to the labour market is a critical concern. Specifically, it will put at risk single-income households headed by low-skilled women, depress total earnings in two-income families, and widen the already-troubling gender gap around the world.

But what about new roles and job categories? What new opportunities could exist for women in a labour market transformed by the fourth industrial revolution? While it is difficult to map the competencies and skills expected in industries not yet created, we can reasonably assume that demand will increase for skills that enable workers to design, build and work alongside technological systems, or in areas that fill the gaps left by these technological innovations.

Because men still tend to dominate computer science, mathematical and engineering professions, increased demand for specialized technical skills may exacerbate gender inequalities. Yet demand may grow for roles that machines cannot fulfil and which rely on intrinsically human traits and capabilities such as empathy and compassion. Women are prevalent in many such occupations including psychologists, therapists, coaches, event planners, nurses and other providers of healthcare.

A key issue here is the relative return on time and effort for roles requiring different technical capabilities, as there is a risk that personal services and other currently female-dominated job categories will remain undervalued. If so, the fourth industrial revolution may lead to further divergence between men's roles and women's. This would be a negative outcome of the fourth industrial revolution, as it would increase both inequality overall and the gender gap, making it more difficult for women to leverage their talents in the workforce of the future. It would also put at risk the value created by increased diversity and the gains that we know organizations can make from the enhanced creativity and efficiency of having gender-balanced teams at all levels. Many of the traits and capabilities traditionally associated with women and female professions will be much more needed in the era of the fourth industrial revolution.

While we cannot predict the different impact on men and women that the fourth industrial revolution will have, we should take the opportunity of a transforming economy to redesign labour policies and business practices to ensure that both men and women are empowered to their full extent.

In tomorrow's world, many new positions and professions will emerge, driven not only by the fourth industrial revolution, but also by non-technological factors such as demographic pressures, geopolitical shifts and new social and cultural norms. Today, we cannot foresee exactly what these will be but I am convinced that talent, more than capital, will represent the critical production factor. For this reason, scarcity of a skilled workforce rather the availability of capital is more likely to be the crippling limit to innovation, competiveness and growth.

This may give rise to a job market increasingly segregated into low-skill/low-pay and high-skill/high-pay segments, or as author and Silicon Valley software entrepreneur Martin Ford predicts,[25] a hollowing out of the entire base of the job skills pyramid, leading in turn to growing inequality and an increase in social tensions unless we prepare for these changes today.

Such pressures will also force us to reconsider what we mean by "high skill" in the context of the fourth industrial revolution. Traditional definitions of skilled labour rely on the presence of advanced or specialised education and a set of defined capabilities within a profession or domain of expertise. Given the increasing rate of change of technologies, the fourth industrial revolution will demand and place more emphasis on the ability of workers to adapt continuously and learn new skills and approaches within a variety of contexts.

The Forum's *Future of Jobs* study also showed that less than 50% of chief human resources officers are at least reasonably confident in their organization's workforce strategy to prepare for these shifts. The main barriers to a more decisive approach include companies' lack of understanding of the nature of disruptive changes, little or no alignment between workforce strategies and firms' innovation strategies, resource constraints and short-term profitability pressures. As a consequence, there is a mismatch between the magnitude of the upcoming changes and the relatively marginal actions being taken by companies to address these challenges. Organizations require a new mindset to meet their own talent needs and to mitigate undesirable societal outcomes.

Impact on developing economies

It is important to reflect upon what this might mean for developing countries. Past phases of the industrial revolution have not yet reached many of the world's citizens, who still do not have access to electricity, clean water, sanitation and many types of capital equipment taken for granted in advanced economies. Despite this, the fourth industrial revolution will inevitably impact developing economies.

As yet, the precise impact of the fourth industrial revolution remains to be seen. In recent decades, although there has been a rise in inequality within countries, the disparity across countries has decreased significantly. Does the fourth industrial revolution risk reversing the narrowing of the gaps between economies that we have seen to date in terms of income, skills, infrastructure, finance and other areas? Or will technologies and rapid changes be harnessed for development and hasten leapfrogging?

These difficult questions must be given the attention they require, even at a time when the most advanced economies are preoccupied with their own challenges. Ensuring that swathes of the globe are not left behind is not a moral imperative; it is a critical goal that would mitigate the risk of global instability due to geopolitical and security challenges such as migration flows.

One challenging scenario for low-income countries is if the fourth industrial revolution leads to significant "re-shoring" of global manufacturing to advanced economies, something very possible if access to low-cost labour no longer drives the competitiveness of firms. The ability to develop strong manufacturing sectors serving the global economy based on cost advantages is a well-worn development pathway, allowing countries to accumulate capital, transfer technology and raise incomes. If this pathway closes, many countries will have to rethink their models and strategies of industrialization. Whether and how developing economies can leverage the opportunities of the fourth industrial revolution is a matter of profound importance to the world; it is essential that further research and thinking be undertaken to understand, develop and adapt the strategies required.

The danger is that the fourth industrial revolution would mean that a winner-takes-all dynamic plays out between countries as well as within them. This would further increase social tensions and conflicts, and create a less cohesive, more volatile world, particularly given that people are today much more aware of and sensitive to social injustices and the discrepancies in living conditions between different countries. Unless public- and private-sector leaders assure citizens that they are executing credible strategies to improve peoples' lives, social unrest, mass migration, and violent extremism could intensify, thus creating risks for countries at all stages of development. It is crucial that people are secure in the belief that they can engage in meaningful work to support themselves and their families, but what happens if there is insufficient demand for labour, or if the skills available no longer match the demand?

3.1.3 The Nature of Work

The emergence of a world where the dominant work paradigm is a series of transactions between a worker and a company more than an enduring relationship was described by Daniel Pink 15 years ago in his book *Free Agent Nation*.[26] This trend has been greatly accelerated by technological innovation.

Today, the on-demand economy is fundamentally altering our relationship with work and the social fabric in which it is embedded. More employers are using the "human cloud" to get things done. Professional activities are dissected into precise assignments and discrete projects and then thrown into a virtual cloud of aspiring workers located anywhere in the world. This is the new on-demand economy, where providers of labour are no longer employees in the traditional sense but rather independent workers who perform specific tasks. As Arun Sundararajan, professor at the Stern School of Business at New York University (NYU), put it in a *New York Times* column by journalist Farhad Manjoo: "We may end up with a future in which a fraction of the workforce will do a portfolio of things to generate an income – you could be an Uber driver, an Instacart shopper, an Airbnb host and a Taskrabbit".[27]

The advantages for companies and particularly fast-growing start-ups in the digital economy are clear. As human cloud platforms classify workers as self-employed, they are – for the moment – free of the requirement to pay minimum wages, employer taxes and social benefits. As explained by Daniel Callaghan, chief executive of MBA & Company in the UK, in a *Financial Times* article: "You can now get whoever you want, whenever you want, exactly how you want it. And because they're not employees you don't have to deal with employment hassles and regulations."[28]

For the people who are in the cloud, the main advantages reside in the freedom (to work or not) and the unrivalled mobility that they enjoy by belonging to a global virtual network. Some independent workers see this as offering the ideal combination of a lot of freedom, less stress and greater job satisfaction. Although the human cloud is in its infancy, there is already substantial anecdotal evidence that it entails silent offshoring (silent because human cloud platforms are not listed and do not have to disclose their data).

Is this the beginning of a new and flexible work revolution that will empower any individual who has an internet connection and that will eliminate the shortage of skills? Or will it trigger the onset of an inexorable race to the bottom in a world of unregulated virtual sweatshops? If the result is the latter – a world of the precariat, a social class of workers who move from task to task to make ends meet while suffering a loss of labour rights, bargaining rights and job security – would this create a potent source of social unrest and political instability? Finally, could the development of the human cloud merely accelerate the automation of human jobs?

The challenge we face is to come up with new forms of social and employment contracts that suit the changing workforce and the evolving nature of work. We must limit the downside of the human cloud in terms of possible exploitation, while neither curtailing the growth of the labour market nor preventing people from working in the manner they choose. If we are unable to do this, the fourth industrial revolution could lead to the dark side of the future of work, which Lynda Gratton, a professor of management practice at London Business School describes in her book *The Shift: The Future of Work is Already Here* - increasing levels of fragmentation, isolation and exclusion across societies.[29]

As I state throughout this book, the choice is ours. It entirely depends on the policy and institutional decisions we make. One has to be aware, however, that a regulatory backlash could happen, thereby reasserting the power of policymakers in the process and straining the adaptive forces of a complex system.

The importance of purpose

We must also keep in mind that it is not only about talent and skills. Technology enables greater efficiency, which most people want. Yet they also wish to feel that they are not merely part of a process but of something bigger than themselves. Karl Marx expressed his concern that the process of specialization would reduce the sense of purpose that we all seek from work, while Buckminster

Fuller cautioned that the risks of over-specialization tend "to shut off the wide-band tuning searches and thus to preclude further discovery of the all-powerful generalized principles."[30]

Now, faced with a combination of increased complexity and hyper-specialization, we are at a point where the desire for purposeful engagement is becoming a major issue. This is particularly the case for the younger generation who often feel that corporate jobs constrain their ability to find meaning and purpose in life. In a world where boundaries are disappearing and aspirations are changing, people want not only work-life balance but also harmonious work-life integration. I am concerned that the future of work will only allow a minority of individuals to achieve such fulfilment.

3.2 Business

Beyond the changes in growth patterns, labour markets and the future of work that will naturally influence all organisations, there is evidence that the technologies that underpin the fourth industrial revolution are having a major impact on how businesses are led, organized and resourced. One particular symptom of this phenomenon is that the historical reduction in the average lifespan of a corporation listed on the S&P 500 has dropped from around 60 to approximately 18.[31] Another is the shift in the time it takes new entrants to dominate markets and hit significant revenue milestones. Facebook took six years to reach revenue of $1 billion a year, and Google just five years. There is no doubt that emerging technologies, almost always powered and enabled by digital capabilities, are increasing the speed and scale of change for businesses.

This also reinforces an underlying theme in my conversations with global CEOs and senior business executives; namely, that the deluge of information available today, the velocity of disruption and the acceleration of innovation are hard to comprehend or anticipate. They constitute a source of constant surprise. In such a context, it is a leader's ability to continually learn, adapt and challenge his or her own conceptual and operating models of success that will distinguish the next generation of successful business leaders.

Therefore, the first imperative of the business impact made by the fourth industrial revolution is the urgent need to look at oneself as a business leader and at one's own organization. Is there evidence of the organization and leadership capacity to learn and change? Is there a track record of prototyping and investment decision-making at a fast pace? Does the culture accept innovation and failure? Everything I see indicates that the ride will only get faster, the changes will be fundamental, and the journey will therefore require a hard and honest look at the ability of organizations to operate with speed and agility.

Sources of disruption

Multiple sources of disruption trigger different forms of business impact. On the supply side, many industries are seeing the introduction of new technologies that create entirely new ways of serving existing needs and significantly disrupt existing value chains. Examples abound. New storage and grid technologies in energy will accelerate the shift towards more decentralized sources. The widespread adoption of 3D printing will make distributed manufacturing and spare-part maintenance easier and cheaper. Real-time information and intelligence will provide unique insights on customers and asset performance that will amplify other technological trends.

Disruption also flows from agile, innovative competitors who, by accessing global digital platforms for research, development, marketing, sales and distribution, can overtake well established incumbents faster than ever by improving the quality, speed or price at which they deliver value. This is the reason why many business leaders consider their biggest threat to be competitors that are not yet regarded as such. It would be a mistake, however, to think that competitive disruption will come only through start-ups. Digitization also enables large incumbents to cross industry boundaries by leveraging their customer base, infrastructure or technology. The move of telecommunications companies into healthcare and automotive segments are examples. Size can still be a competitive advantage if smartly leveraged.

Major shifts on the demand side are also disrupting business: Increasing transparency, consumer engagement and new patterns of consumer behaviour (increasingly built upon access to mobile networks and data) force companies to adapt the way they design, market and deliver existing and new products and services.

Overall, I see the impact of the fourth industrial revolution on business as an inexorable shift from the simple digitization that characterized the third industrial revolution to a much more complex form of innovation based on the combination of multiple technologies in novel ways. This is forcing all companies to re-examine the way they do business and takes different forms. For some companies, capturing new frontiers of value may consist of developing new businesses in adjacent segments, while for others, it is about identifying shifting pockets of value in existing sectors.

The bottom line, however, remains the same. Business leaders and senior executives need to understand that disruption affects both the demand and supply sides of their business. This, in turn, must compel them to challenge the assumptions of their operating teams and find new ways of doing things. In short, they have to innovate continuously.

Four major impacts

The fourth industrial revolution has four main effects on business across industries:

– customer expectations are shifting
– products are being enhanced by data, which improves asset productivity
– new partnerships are being formed as companies learn the importance of new forms of collaboration, and
– operating models are being transformed into new digital models.

3.2.1 Customer Expectations

Customers, whether as individuals (B2C) or businesses (B2B), are increasingly at the centre of the digital economy, which is all about how they are served. Customer expectations are being redefined into experiences. The Apple experience, for example, is not just about how we use the product but also about the packaging, the brand, the shopping and the customer service. Apple is thus redefining expectations to include product experience.

Traditional approaches to demographic segmentation are shifting to targeting through digital criteria, where potential customers can be identified based on their willingness to share data and interact. As the shift from ownership to shared access accelerates (particularly in cities), data sharing will be a necessary part of the value proposition. For example, car-sharing schemes will require the integration of personal and financial information across multiple companies in the automotive, utility, communications and banking sectors.

Most companies profess to be customer-centric, but their claims will be tested as real-time data and analytics are applied to the way they target and serve their customers. The digital age is about accessing and using data, refining the products and experiences, and moving to a world of continual adjustment and refinement while ensuring that the human dimension of the interaction remains at the heart of the process.

It is the ability to tap into multiple sources of data – from personal to industrial, from lifestyle to behavioural – that offers granular insights into the customer's purchasing journey that would have been inconceivable until recently. Today, data and metrics deliver in quasi-real time critical insights into customer needs and behaviours that drive marketing and sales decisions.

This trend of digitization is currently towards more transparency, meaning more data in the supply chain, more data at the fingertips of consumers and hence more peer-to-peer comparisons on the performance of products that shift power to consumers. As an example, price-comparison websites make it easy to compare prices, the quality of service, and the performance of the product. In a mouse click or finger swipe, consumers instantaneously move away from one brand, service or digital retailer to the next. Companies are no longer able to shirk accountability for poor performance. Brand equity is a prize hard won and easily lost. This will only be amplified in a more transparent world.

To a large extent, the millennial generation is setting consumer trends. We now live in an on-demand world where 30 billion WhatsApp messages are sent every day[32] and where 87% of young people in the US say their smart phone never leaves their side and 44% use their camera function daily.[33] This is a world which is much more about peer-to-peer sharing and user-generated content. It is a world of the *now*: a real-time world where traffic directions are instantly provided and groceries are delivered directly to your door. This "now world" requires companies to respond in real time wherever they are or their customers or clients may be.

It would be a mistake to assume that this is confined to high-income economies. Take online shopping in China. On 11 November 2015, dubbed Singles Day by the Alibaba Group, the e-commerce service handled online transactions worth more than $14 billion, with 68% of sales through mobile devices.[34] Another example is sub-Saharan Africa, which is the fastest-growing region in terms of mobile-phone subscriptions, demonstrating how mobile internet is leapfrogging fixed-line access. GSM Association expects an additional 240 million mobile internet users in sub-Saharan Africa over the next five years.[35] And while advanced economies have the highest penetration rates of social media, East Asia, South-East Asia and Central America are above the global average of 30% and growing fast. WeChat (Weixin), a China-based mobile text and voice messaging service, gained around 150 million users in just 12 months to the end of 2015, year-on-year growth of at least 39%.[36]

3.2.2 Data-Enhanced Products

New technologies are transforming how organizations perceive and manage their assets, as products and services are enhanced with digital capabilities that increase their value. Tesla, for example, shows how over-the-air software updates and connectivity can be used to enhance a product (a car) after purchase, rather than let it depreciate over time.

Not only are new materials making assets more durable and resilient but data and analytics are also transforming the role of maintenance. Analysis provided by sensors placed on assets enables their constant monitoring and proactive maintenance and, in doing so, maximizes their utilization. It is no longer about finding specific faults but rather about using performance benchmarks (based on data supplied by sensors and monitored through algorithms) that can highlight when a piece of equipment is moving outside its normal operating window. On aircrafts, for example, the airline control centres know before the pilots do if an engine is developing a fault on a particular plane. They can therefore instruct the pilot on what to do and mobilize the maintenance crew in advance at the flight destination.

In addition to maintenance, the ability to predict the performance of an asset enables new business models to be established. Asset performance can be measured and monitored over time – analytics provide insights on operational tolerances and provide the basis for outsourcing products that are not core or strategic to the needs of the business. SAP is an example of a company that is leveraging data from physical products embedded in agriculture to increase uptime and utilization.

The ability to predict the performance of an asset also offers new opportunities to price services. Assets with high throughput such as lifts or walkways can be priced by asset performance, and service providers can be paid on the basis of actual performance against a threshold of 99.5% uptime over a given period. Take the example of truck fleets. Long-distance haulers are interested in propositions where they pay tire manufacturers by the 1,000 kilometres of road use rather than periodically buying new tires. This is because the combination of sensors and analytics enables tire companies to monitor driver performance, fuel consumption and tire wear to offer a complete end-to-end service.

3.2.3 Collaborative Innovation

A world of customer experiences, data-based services and asset performance through analytics requires new forms of collaboration, particularly given the speed at which innovation and disruption are taking place. This is true for incumbents and established businesses but also for young, dynamic firms. The former often lack specific skills and have lower sensitivity to evolving customer needs, while the latter are capital poor and lack the rich data generated by mature operations.

As a the Forum's *Collaborative Innovation: Transforming Business, Driving Growth* report outlines, when firms share resources through collaborative innovation, significant value can be created for both parties as well as for the economies in which such collaborations take place. One such example is the recent collaboration between the industrial giant Siemens, which spends around $4 billion a year in research and development, and Ayasdi, an innovative machine-learning company and Forum Technology Pioneer founded at Stanford University in 2008. This partnership gives Siemens access to a partner that can help solve complex challenges of extracting insights from vast data, while Ayasdi can validate its topological data analysis approach with real-world data, while expanding market presence.

Such collaborations, however, are often far from straightforward. They require significant investment from both parties to develop firm strategy, search for appropriate partners, establish communication channels, align processes, and flexibly respond to changing conditions, both inside and outside the partnership. Sometimes, such collaborations spawn entirely new business models such as city car-sharing schemes, which bring together businesses from multiple industries to provide an integrated customer experience. This is only as good as the weakest link in the partnership chain. Companies need to go well beyond marketing and sales agreements to understand how to adopt comprehensive collaborative approaches. The fourth industrial revolution forces companies to think about how offline and online worlds work together in practice.

3.2.4 New Operating Models

All these different impacts require companies to rethink their operating models. Accordingly, strategic planning is being challenged by the need for companies to operate faster and with greater agility.

As mentioned earlier, an important operating model enabled by the network effects of digitization is the platform. While the third industrial revolution saw the emergence of purely digital platforms, a hallmark of the fourth industrial revolution is the appearance of global platforms intimately connected to the physical world. The platform strategy is both profitable and disruptive. Research by the MIT Sloan School of Management shows that 14 out of the top 30 brands by market capitalization in 2013 were platform-oriented companies.[37]

Platform strategies, combined with the need to be more customer-centric and to enhance products with data, are shifting many industries from a focus on selling products to delivering services. An increasing number of consumers no longer purchase and own physical objects, but rather pay for the delivery of the underlying service which they access via a digital platform. It is possible, for example, to get digital access to billions of books via Amazon's Kindle Store, to play almost any song in the world via Spotify, or to join a car-sharing enterprise that provides mobility services without the need to own the vehicle. This shift is a powerful one and allows for more transparent, sustainable models of exchanging value in the economy. But it also creates challenges in how we define ownership, how we curate and engage with unlimited content, and how we interact with the increasingly-powerful platforms that provide these services at scale.

The World Economic Forum's work in its Digital Transformation of Industry initiative highlights a number of other business and operating models designed to capitalize on the fourth industrial revolution. The previously mentioned "customer-centricity" is one of these, with proponents such as Nespresso focusing their efforts on front-line processes and empowering staff to put the client first. Frugal business models use the opportunities afforded by the interaction of digital, physical and human realms to open up new forms of optimization such as efforts by Michelin to provide high-quality services at low cost.

Data-powered business models create new revenue sources from their access to valuable information on customers in a broader context and increasingly rely on analytics and software intelligence to unlock insights. "Open and liquid" companies position themselves as part of a fluid ecosystem of value creation, while "Skynet" firms focus on automation, becoming more prevalent in hazardous industries and locations. And there are many examples of businesses pivoting towards business models that focus on employing new technologies to make more efficient use of energy and material flows, thereby preserving resources, lowering costs, and having a positive impact on the environment (see Box B: Environmental Renewal and Preservation).

These transformations mean that businesses will need to invest heavily in cyber- and data-security systems to avoid direct disruption by criminals, activists or unintentional failures in digital infrastructure. Estimates of the total annual cost to business of cyber-attacks are of the order of magnitude of $500 billion. The experiences of companies such as Sony Pictures, TalkTalk, Target and Barclays indicate that losing control of sensitive corporate and customer data has a material negative effect on share prices. This accounts for why Bank of America Merrill Lynch estimates that the cyber-security market will more than double from around $75 billion in 2015 to $170 billion by 2020, implying an annual growth rate of more than 15% for the industry in the coming five years.[38]

Emerging operating models also mean that talent and culture have to be rethought in light of new skill requirements and the need to attract and retain the right sort of human capital. As data become central to both decision-making and operating models across industries, workforces require new skills, while processes need to be upgraded (for example, to take advantage of the availability of real-time information) and cultures need to evolve.

As I mentioned, companies need to adapt to the concept of "talentism". This is one of the most important, emerging drivers of competitiveness. In a world where talent is the dominant form of strategic advantage, the nature of organizational structures will have to be rethought. Flexible hierarchies, new ways of measuring and rewarding performance, new strategies for attracting and retaining skilled talent will all become key for organizational success. A capacity for agility will be as much about employee motivation and communication as it will be about setting business priorities and managing physical assets.

My sense is that successful organizations will increasingly shift from hierarchical structures to more networked and collaborative models. Motivation will be increasingly intrinsic, driven by the collaborative desire of employees and management for mastery, independence and meaning. This suggests that businesses will become increasingly organized around distributed teams, remote workers and dynamic collectives, with a continuous exchange of data and insights about the things or tasks being worked on.

An emerging workplace scenario that reflects this change builds on the rapid rise of wearable technology when combined with the internet of things, which is progressively enabling companies to blend digital and physical experiences to benefit workers as well as consumers. For example, workers operating with highly complex equipment or in difficult situations can use wearables to help design and repair components. Downloads and updates to connected machinery ensure that both workers in the field and the capital equipment they use are kept up to date with the latest developments. In the world of the fourth industrial revolution, where it is standard practice to upgrade cloud-based software and refresh data assets through the cloud, it will be even more important to ensure that humans and their skills keep pace.

Combining the digital, physical and biological worlds

Companies able to combine multiple dimensions – digital, physical and biological – often succeed in disrupting an entire industry and their related systems of production, distribution and consumption.

Uber's popularity in many cities starts with an improved customer experience – tracking of the car location via a mobile device, a description of the car standards and a seamless payment process, thus avoiding delays at the destination. The experience has been enhanced and bundled with the physical product (transportation of a person from A to B) by optimizing the utilization of the asset (the car owned by the driver). In such cases, the digital opportunities are often not translated into just a higher price or lower cost but also into a fundamental change of the business model. This is driven by an end-to-end approach, from service acquisition to delivery.

These combination-based business models illustrate the extent of the disruption that occurs when digital assets and interesting combinations of existing digital platforms are used to reorganize relationships with physical assets (marking a notable shift from ownership to access). In their markets, neither company owns the assets: a car driver owns the car and makes it available; a homeowner makes his room available. In both cases, the competitive advantage is built on a superior experience, combined with reduced transaction and friction costs. Also, these companies match demand and supply in a rapid and convenient manner, which side steps the business models of the incumbents.

This marketplace approach progressively erodes the long established position of incumbents and dismantles the boundaries between industries. Many senior executives expect industry convergence to be the primary force impacting their business in the next three to five years.[39] Once a customer has established a track record of trust and confidence on the platform, it becomes easy for the digital provider to offer other products and services.

Fast-moving competitors provoke a disaggregation of the more traditional industry silos and value chains, and also disintermediate the existing relationship between businesses and their customers. New disruptors can rapidly scale at a much lower cost than the incumbents, generating in the process a rapid growth in their financial returns through network effects. Amazon's evolution from a bookseller to a $100 billion a year retail conglomerate shows how customer loyalty, combined with insights on preferences and solid execution can lead to selling across multiple industries. It also demonstrates the benefits of scale.

In almost all industries, digital technologies have created new, disruptive ways of combining products and services – and in the process, have dissolved the traditional boundaries between industries. In the automotive realm, a car is now a computer on wheels, with electronics representing roughly 40% of the cost of a car. The decision by Apple and Google to enter the automotive market shows that a tech company can now transform into a car company. In the future, as the value shifts towards the electronics, the technology and licensing software may prove more strategically beneficial than manufacturing the car per se.

The finance industry is going through a similar period of disruptive change. P2P (peer-to-peer) platforms are now dismantling barriers to entry and lowering costs. In the investment business, new "robo-advisory" algorithms and their corresponding apps provide advisory services and portfolio tools at a fraction of the old transaction cost – 0.5% instead of the traditional 2%, thereby threatening a whole segment of the current financial industry. The industry is also aware that blockchain will soon revolutionize the way it operates because its possible applications in finance have the opportunity to reduce settlement and transaction costs by up to $20 billion and transform the way the industry works. The shared database technology can streamline such varied activities as the storage of clients' accounts, cross-border payments, and the clearing and settling of trades, as well as products and services that do not exist yet, such as smart futures contracts that self-execute without a trader (e.g. a credit derivative that pays out automatically when a country or company defaults).

The healthcare industry is also faced with the challenge of incorporating simultaneous advances in physical, biological and digital technologies, as the development of new diagnostic approaches and therapies coincide with a push to digitize patient records and capitalize on the wealth of information able to be gathered from wearable devices and implantable technologies.

Not all industries are at the same point of disruption, but all are being pushed up a curve of transformation by the forces driving the fourth industrial revolution. There are differences according to industry and demographic profile of the customer base. But in a world characterized by uncertainty, the ability to adapt is critical – if a company is unable to move up the curve, it may be pushed off it.

The companies that survive or thrive will need to maintain and continually sharpen their innovative edge. Businesses, industries and corporations will face continuous Darwinian pressures and as such, the philosophy of "always in beta" (always evolving) will become more prevalent. This suggests that the global number of entrepreneurs and intrapreneurs (enterprising company managers) will increase. Small and medium-sized enterprises will have the advantages of speed and the agility needed to deal with disruption and innovation.

Large organizations, by contrast, will survive by leveraging their scale advantages and investing in their ecosystem of start-ups and SMEs by acquiring and partnering with smaller and more innovative businesses. This will enable them to maintain autonomy in their respective businesses while also allowing for more efficient and agile operations. Google's recent decision to reorganize into a holding company called Alphabet is a vivid example of this trend, driven by the need to sustain its innovative character and maintain its agility.

Finally, as the next sections detail, the regulatory and legislative landscapes will significantly shape how researchers, businesses and citizens develop, invest in and adopt both emerging technologies and the operating models that enable them to create value for users. While new technologies and innovative businesses offer new products and services that can improve the lives of many, those same technologies and the systems that support them could also create impacts we wish to avoid. These range from widespread unemployment and increased inequality, which was discussed previously, to the dangers of automated weapons systems and new cyber risks.

While perspectives on what constitutes the right mix of regulation may vary, my conversations with government, business and civil society leaders indicate that they share the same overarching goal: to create agile, responsible regulatory and legislative ecosystems that will allow innovation to thrive while minimizing its risks to ensure the stability and prosperity of society.

Box B: Environmental Renewal and Preservation

The convergence of the physical, digital and biological worlds that is at the heart of the fourth industrial revolution offers significant opportunities for the world to achieve huge gains in resource use and efficiency. As Project MainStream, the World Economic Forum's initiative to accelerate the transition to the circular economy, has shown, the promise is not just that individuals, organizations and governments can have less impact on the natural world but also that there is great potential to restore and regenerate our natural environment through the use of technologies and intelligent systems design.

At the heart of this promise is the opportunity to shift businesses and consumers away from the linear take-make-dispose model of resource use, which relies on large quantities of easily accessible resources, and towards a new industrial model where effective flows of materials, energy, labour and now information interact with each other and promote by design a restorative, regenerative and more productive economic system.

There are four pathways that help take us there. First, thanks to the internet of things (IoT) and intelligent assets, it is now possible to track materials and energy flows so as to achieve huge new efficiencies all the way along value chains. Of the $14.4 trillion in economic benefits that Cisco estimates will be realized from the IoT in the next decade, $2.7 trillion in value can be gained from elimination of waste and improved processes in supply chains and logistics. IoT-enabled solutions could reduce greenhouse gas emissions by 9.1 billion tonnes by 2020, representing 16.5% of the projected total in that year.[40]

Second, the democratization of information and transparency that comes from digitized assets gives new powers to citizens to hold companies and countries accountable. Technologies such as blockchain will help make this information more trustworthy, for example by capturing and certifying satellite monitoring data on deforestation in a secure format to hold landholders to account more closely.

Third, new information flows and increasing transparency can help shift citizen behaviour on a large scale, as it becomes the path of least resistance within a new set of business and social norms for a sustainable circular system. Fruitful convergence between the fields of economics and psychology has been producing insights into how we perceive the world, behave and justify our behaviour, while a number of large-scale randomized control trials by governments, corporations and universities have shown that this can work. One example is OPower, which uses peer-comparison to entice people into consuming less electricity, thereby protecting the environment while reducing costs.

Fourth, as the previous section detailed, new business and organizational models promise innovative ways of creating and sharing value, which in turn lead to whole system changes that can actively benefit the natural world as much as our economies and societies. Self-driving vehicles, the sharing economy and leasing models all result in significantly higher asset utilization rates, as well as making it far easier to capture, reuse and "upcycle" materials when the appropriate time comes.

The fourth industrial revolution will enable firms to extend the use-cycle of assets and resources, increase their utilization and create cascades that recover and repurpose materials and energy for further uses, lowering emissions and resource loads in the process. In this revolutionary new industrial system, carbon dioxide turns from a greenhouse pollutant into an asset, and the economics of carbon capture and storage move from being cost as well as pollution sinks to becoming profitable carbon-capture and use-production facilities. Even more importantly, it will help companies, governments and citizens become more aware of and engaged with strategies to actively regenerate natural capital, allowing intelligent and regenerative uses of natural capital to guide sustainable production and consumption and give space for biodiversity to recover in threatened areas.

3.3 National and Global

The disruptive changes brought by the fourth industrial revolution are redefining how public institutions and organizations operate. In particular, they compel governments – at the regional, national and local levels - to adapt by reinventing themselves and by finding new ways of collaboration with their citizens and the private sector. They also affect how countries and governments relate to each other.

In this section, I explore the role that governments must assume to master the fourth industrial revolution, while recognizing the enduring forces that are changing the traditional perceptions of politicians and their role in society. With growing citizen empowerment and greater fragmentation and polarization of populations, this could result in political systems that make governing more difficult and governments less effective. This is particularly important as it occurs at a time when governments should be essential partners in shaping the transition to new scientific, technological, economic and societal frameworks.

3.3.1 Governments

When assessing the impact of the fourth industrial revolution on governments, the use of digital technologies to govern better is top-of-mind. More intense and innovative use of web technologies can help public administrations modernize their structures and functions to improve overall performance, from strengthening processes of e-governance to fostering greater transparency, accountability and engagement between the government and its citizens. Governments must also adapt to the fact that power is also shifting from state to non-state actors, and from established institutions to loose networks. New technologies and the social groupings and interactions they foster allow virtually anyone to exercise influence in a way that would have been inconceivable just a few years ago.

Governments are among the most impacted by this increasingly transient and evanescent nature of power. As Moisés Naím puts it, "in the 21st century, power is easier to get, harder to use, and easier to lose."[41] There is little doubt that governing is tougher today than in the past. With a few exceptions, policymakers are finding it harder to effect change. They are constrained by rival power centres including the transnational, provincial, local and even the individual. Micro-powers are now capable of constraining macro-powers such as national governments.

The digital age undermined many of the barriers that used to protect public authority, rendering governments much less efficient or effective as the governed, or the public, became better informed and increasingly demanding in their expectations. The WikiLeaks saga – in which a tiny non-state entity confronted a mammoth state – illustrates the asymmetry of the new power paradigm and the erosion of trust that often comes with it.

It would take a book dedicated to the subject alone to explore all the multifaceted impacts of the fourth industrial revolution on governments, but the key point is this: Technology will increasingly enable citizens, providing a new way to voice their opinions, coordinate their efforts and possibly circumvent government supervision. I say "possibly", because the opposite might just as well be true, with new surveillance technologies giving rise to all-too-powerful public authorities.

Parallel structures will be able to broadcast ideologies, recruit followers and coordinate actions against – or in spite of – official governmental systems. Governments, in their current form, will be forced to change as their central role of conducting policy increasingly diminishes due to the growing levels of competition and the redistribution and decentralization of power that new technologies make possible. Increasingly, governments will be seen as public-service centres that are evaluated on their abilities to deliver the expanded service in the most efficient and individualized ways.

Ultimately, it is the ability of governments to adapt that will determine their survival. If they embrace a world of exponentially disruptive change, and if they subject their structures to the levels of transparency and efficiency that can help them maintain their competitive edge, they will endure. In doing so, however, they will be completely transformed into much leaner and more efficient power cells, all within an environment of new and competing power structures.

As in previous industrial revolutions, regulation will play a decisive role in the adaptation and diffusion of new technologies. However, governments will be forced to change their approach when it comes to the creation, revision and enforcement of regulation. In the "old world", decision-makers had enough time to study a specific issue and then create the necessary response or appropriate regulatory framework. The whole process tended to be linear and mechanistic, following a strict top-down approach. For a variety of reasons, this is no longer possible.

With the rapid pace of change triggered by the fourth industrial revolution, regulators are being challenged to an unprecedented degree. Today's political, legislative and regulatory authorities are often overtaken by events, unable to cope with the speed of technological change and the significance of its implications. The 24-hour news cycle puts pressure on leaders to comment or act immediately to events, reducing the time available for arriving at measured, principled and calibrated responses. There is a real danger of loss-of-control over what matters, particularly in a global system with almost 200 independent states and thousands of different cultures and languages.

In such conditions, how can policymakers and regulators support technological developments without stifling innovation while preserving the interest of the consumers and the public at large? Agile governance is the response (see Box C: Agile Governance Principles in an Age of Disruption).

Many of the technological advances we currently see are not properly accounted for in the current regulatory framework and might even disrupt the social contract that governments have established with their citizens. Agile governance means that regulators must find ways to adapt continuously to a new, fast-changing environment by reinventing themselves to understand better what they are regulating. To do so, governments and regulatory agencies need to closely collaborate with business and civil society to shape the necessary global, regional and industrial transformations.

Agile governance does not imply regulatory uncertainty, nor frenetic, ceaseless activity on the part of policymakers. We should not make the mistake of thinking that we are caught between two equally unpalatable legislative frameworks – outdated but stable on one hand, or up-to-date but volatile on the other. In the age of the fourth industrial revolution, what is needed is not necessarily more or faster policy-making, but rather a regulatory and legislative ecosystem that can produce more resilient frameworks. This approach could be enhanced by creating more space for stillness in order to reflect on important decisions. The challenge is to make this deliberation far more productive than is currently the case, infused with foresight to create maximum space for innovation to emerge.

In summary, in a world where essential public functions, social communication and personal information migrate to digital platforms, governments – in collaboration with business and civil society – need to create the rules, checks and balances to maintain justice, competitiveness, fairness, inclusive intellectual property, safety and reliability.

Two conceptual approaches exist. In the first, everything that is not explicitly forbidden is allowed. In the second, everything that is not explicitly allowed is forbidden. Governments must blend these approaches. They have to learn to collaborate and adapt, while ensuring that the human being remains at the centre of all decisions. This is the challenge for governments, which have never been more necessary than in this fourth industrial revolution: they must let innovation flourish, while minimizing risks.

To achieve this, governments will need to engage citizens more effectively and conduct policy experiments that allow for learning and adaptation. Both of these tasks mean that governments and citizens alike must rethink their respective roles and how they interact with one another, simultaneously raising expectations while explicitly acknowledging the need to incorporate multiple perspectives and allow for failure and missteps along the way.

Box C: Agile Governance Principles in an Age of Disruption

Job market

Digital technologies and global communication infrastructure significantly change the traditional concepts of work and pay, enabling the emergence of new types of jobs that are extremely flexible and inherently transient (the so-called on-demand economy). While these new jobs allow for people to enjoy more flexible working hours and might unleash a whole new wave of innovation in the job marketplace, they also raise important concerns with regard to the reduced degree of protection in the context of the on-demand economy, where every worker has essentially become a contractor, who no longer benefits from job security and longevity.

Money and taxation

The on-demand economy is also raising serious issues
with regard to tax collection, as it becomes much easier
and attractive for transient workers to operate in the
black market. While digitally mediated payment systems
are making transactions and micro-transactions more
transparent, new decentralized payment systems are
emerging today, which could significantly hinder the ability
for public authorities and private actors to trace the origin
and destination of such transactions.

Liability and protection

Government-issued monopolies (e.g. the taxi industry,
medical practitioners) have long been justified on the
grounds that certain types of high-risk professions require
a higher degree of scrutiny and should only be performed
by licensed professionals so as to ensure a proper degree
of safety and consumer protection. Many of these
government-issued monopolies are now being disrupted by
technological advances which enable people to interact with
one another on a peer-to-peer basis and by the emergence
of novel intermediaries in charge of coordinating peers and
facilitating their interactions.

Security and privacy

Despite the transnational character of the internet network
and the growing global economy, data rights and data
protection regulations are still heavily fragmented. Rules
around the collection, processing and reselling of personal
data are well defined in Europe but are still weak or entirely
lacking in many other jurisdictions. The aggregation of large
datasets is making it possible for large online operators to
deduce more information than was actually provided (either
implicitly or explicitly) by users. User profiling through big-
data analysis and inference techniques is opening the way
for new, much more customized and personalized services,
which can benefit users and consumers, but which also
raise important concerns when it comes to user privacy
and individual autonomy. Given increased concerns around

cyber crime and identity theft, in many jurisdictions, the balance between surveillance and freedom is rapidly tipping towards increased monitoring, as shown by revelations brought to light by Edward Snowden, the American intelligence analyst who leaked documents relating to US national security operations.

Availability and inclusion

As the global economy increasingly moves into the digital realm, the availability of reliable internet infrastructure becomes a crucial prerequisite for a flourishing economy. Governments need to understand the potential provided by these technological advances. Not only do they need to adopt these technologies to optimize their internal operations, they also need to promote and support their widespread deployment and use to move forward towards a globally connected information society. The issue of digital exclusion (or digital divide) becomes ever more pressing, as it is increasingly difficult for people to participate in the digital economy and new forms of civic engagement without proper internet access and/or without access to a connected device or sufficient knowledge to use that device.

Power asymmetries

In today's information society, asymmetries of information might lead to significant asymmetries of power, since whoever has the knowledge to operate the technology also has the power to do so. An entity with root access is almost omnipotent. Given the complexity of fully grasping the potential and underlying technicalities of modern technologies, however, increasing inequalities might emerge between tech-savvy individuals, who understand and control these technologies, and less knowledgeable individuals, who are passive users of a technology they do not understand.

Source: "A call for Agile Governance Principles in an Age of Disruption", Global Agenda Council on Software & Society, World Economic Forum, November 2015

3.3.2 Countries, Regions and Cities

Because digital technology knows no borders, there are many questions that come to mind when considering the geographic impact of technology and the impact of geography on technology. What will define the roles that countries, regions and cities play in the fourth industrial revolution? Will Western Europe and the US lead the transformation, as they did the previous industrial revolutions? Which countries will be able to leapfrog? Will there be greater and more effective collaboration for the bettering of society, or will we see increased fragmentation not only within countries but also across countries? In a world where goods and services can be produced almost everywhere, and where much of the demand for low-skilled and low-wage work is overtaken by automation, will those who can afford it congregate in countries with strong institutions and proven quality of life?

Innovation-Enabling Regulation

In trying to answer these questions, one thing is clear and of great importance: the countries and regions that succeed in establishing tomorrow's preferred international norms in the main categories and fields of the new digital economy (5G communications, the use of commercial drones, the internet of things, digital health, advanced manufacturing and so on) will reap considerable economic and financial benefits. In contrast, countries that promote their own norms and rules to give advantages to their domestic producers, while also blocking foreign competitors and reducing royalties that domestic companies pay for foreign technologies, risk becoming isolated from global norms, putting these nations at risk of becoming the laggards of the new digital economy.[42]

As previously mentioned, the broad issue of legislation and compliance at the national or regional level will play a determining role in shaping the ecosystem in which disruptive companies operate. This sometimes leads countries to lock horns with each other. A good case in point is the October 2015 decision by the European Court of Justice (ECJ) to invalidate the safe-harbour agreement that guided the flow of personal data between the United States and the European Union. This is bound to increase the costs of compliance that companies incur when doing business in Europe and has become a transatlantic issue of contention.

This example reinforces the increasing importance of innovation ecosystems as a key driver of competitiveness. Looking ahead, the distinction between high- and low-cost countries, or between emerging and mature markets, will matter less and less. Instead, the key question will be whether an economy can innovate.

Today, for example, North American companies remain the most innovative in the world by virtually any measure. They attract the top talent, earn the most patents, command the majority of the world's venture capital, and when publicly listed, enjoy high corporate valuations. This is further reinforced by the fact that North America remains at the cutting edge of four synergistic technology revolutions: technology-fuelled innovation in energy production, advanced and digital manufacturing, the life sciences, and information technology.

And while North America and the EU, which includes some of the most innovative economies, lead the way, other parts of the world are rapidly catching up. Estimates of China's innovation performance, for example, have increased to 49% of the EU level in 2015 (up from 35% in 2006) as the country shifts its economic model to focus on innovation and services.[43] Even considering that China's progress springs from a relatively low level, the country is continually entering higher value-added segments of global production and employing its significant economies of scale to compete better globally.[44]

Overall, this shows that policy choices will ultimately determine whether a specific country or region can capitalize fully on the opportunities afforded by the technology revolution.

Regions and cities as hubs of innovation

I am particularly concerned about the effect that automation will have on some countries and regions, particularly those in fast-growing markets and developing countries, where it could abruptly erode the comparative advantage they enjoy in producing labour-intensive goods and services. Such a scenario could devastate the economies of some countries and regions that are currently thriving.

It is clear that neither countries nor regions can flourish if their cities (innovation ecosystems) are not being continually nourished. Cities have been the engines of economic growth, prosperity and social progress throughout history, and will be essential to the future competitiveness of nations and regions. Today, more than half of the world's population lives in urban areas, ranging from mid-size cities to megacities, and the number of city dwellers worldwide keeps rising. Many factors that affect the competitiveness of countries and regions – from innovation and education to infrastructure and public administration – are under the purview of cities.

The speed and breadth by which cities absorb and deploy technology, supported by agile policy frameworks, will determine their ability to compete in attracting talent. Possessing a superfast broadband, putting into place digital technologies in transportation, energy consumption, waste recycling and so on help make a city more efficient and liveable, and therefore more attractive than others.

It is therefore critical that cities and countries around the world focus on ensuring access to and use of the information and communication technologies on which much of the fourth industrial revolution depends. Unfortunately, as the World Economic Forum's *Global Information Technology Report 2015* points out, ICT infrastructures are neither as prevalent nor diffusing as fast as many people believe. "Half of the world's population does not have mobile phones and 450 million people still live out of reach of a mobile signal. Some 90% of the population of low-income countries and over 60% globally are not online yet. Finally, most mobile phones are of an older generation."[45]

Governments must therefore focus on bridging the digital divide in countries at all stages of development to ensure that cities and countries have the basic infrastructure required to create the economic opportunities and shared prosperity that is possible through new models of collaboration, efficiency and entrepreneurship.

The Forum's work on *Data-Driven Development* highlights that it is not just access to digital infrastructure that matters for grasping these opportunities. Also critical is addressing the "data deficit" in many countries, particularly in the global South, given constraints on how data can be created, collected, transmitted and used. Closing the four "gaps" which contribute to this deficit – its existence, access, governance and usability – gives countries, regions and cities many additional abilities that can enhance their development, such as tracking the outbreak of infectious diseases, responding better to natural disasters, enhancing access to public and financial services for the poor, and understanding migration patterns of vulnerable populations.[46]

Countries, regions and cities can do more than simply change the regulatory environment. They can actively invest in becoming launch pads for digital transformation, so as to attract and encourage entrepreneurs and investors in innovative startups while also ensuring that established businesses orient themselves to the opportunities of the fourth industrial revolution. As young, dynamic firms and established enterprises connect with one another and to citizens and universities, cities become both sites of experimentation and powerful hubs for turning new ideas into real value for the local and global economies.

According to the innovation charity Nesta in the UK, the five cities that are globally best placed in terms of having the most effective policy environment to foster innovation are: New York, London, Helsinki, Barcelona and Amsterdam.[47] Nesta's study shows that these cities particularly succeed in finding creative ways to effect change outside of the formal policy arena, being open by default, and acting more like entrepreneurs (than bureaucrats). All three criteria give rise to the best-in-class examples we currently see globally, and which are equally applicable to cities in emerging markets and the developing world. Medellin, Colombia, was honored with a City of the Year award in 2013, recognising its innovative approaches to mobility and environmental sustainability, beating the other finalists New York and Tel Aviv.[48]

In October 2015, the World Economic Forum's Global Agenda Council on the Future of Cities released a report highlighting instances of cities around the world pursuing innovative solutions to a variety of problems (see Box D: Urban Innovations).[49] This work indicates that the fourth industrial revolution is unique, driven as it is by a global network of smart (network-driven) cities, countries and regional clusters, which understand and leverage the opportunities of this revolution – top down and bottom up – acting from a holistic and integrated perspective.

Box D: Urban Innovations

Digitally reprogrammable space: Buildings will be able to instantly shift purpose to serve as a theatre, gymnasium, social centre, nightclub or whatever, thus minimizing the overall urban footprint. This would allow cities to get more from less.

"Waternet": The internet of pipes, this will employ sensors in the water system to monitor flows and thus manage the entire cycle, providing sustainable water for human and ecological needs.

Adopting a tree through social networks: Studies show that increasing a city's green area by 10% could compensate for the temperature increase caused by climate change: vegetation helps to block shortwave radiation while also evaporating water, cooling the ambient air and creating more comfortable microclimates. Tree canopies and root systems can also reduce storm water flows and balance nutrient loads.

Next-generation mobility: With advances in sensors, optics and embedded processors, improved safety for pedestrians and non-motorized transportation will lead to greater adoption of public transport, reduced congestion and pollution, better health and commutes that are quicker, more predictable and less expensive.

Co-generation, co-heating and co-cooling: Co-generation mechanical systems already capture and use the excess heat, significantly improving energy efficiency. Trigeneration systems use the heat either to warm buildings or to cool them through absorption refrigerator technology – for example, cooling office complexes that house large numbers of computers.

Mobility-on-demand: Digitization is making vehicular traffic more efficient by allowing real-time information and an unprecedented monitoring of urban mobility infrastructure. This opens up new potential for leveraging unused vehicle capacity through dynamic optimization algorithms.

Intelligent street poles: Next-generation LED street lights can act as a platform for a host of sensing technologies that collect data on weather, pollution, seismic activity, the movement of traffic and people, and noise and air pollution. By linking these intelligent street poles in a network, it is possible to sense what is going on across a city in real time and provide innovative solutions in areas such as public safety or identifying where there are free parking spaces.

Source: "Top Ten Urban Innovations", Global Agenda Council on the Future of Cities, World Economic Forum, October 2015

3.3.3 International Security

The fourth industrial revolution will have a profound impact on the nature of state relationships and international security. I devote particular attention to this issue in this section as I feel that of all the important transformations linked to the fourth industrial revolution, security is a topic not sufficiently discussed in the public domain and in sectors outside governments and the defence industry.

The critical danger is that a hyperconnected world of rising inequality may lead to increasing fragmentation, segregation and social unrest, which in turn creates the conditions for violent extremism. The fourth industrial revolution will change the character of security threats while also influencing shifts of power, which are occurring both geographically, and from state to non-state actors. Faced with the rise of armed non-state actors within what is already an increasing complex geo-political landscape, the prospect of establishing a common platform for collaboration around key international security challenges becomes a critical, if more demanding challenge.

Connectivity, fragmentation and social unrest

We live in a hyper-connected world, where information, ideas and people are travelling faster than ever before. We also live in a world of rising inequality, a phenomenon that will be exacerbated by the massive changes in the labour market that I described earlier. Widening social exclusion, the challenge of finding reliable sources of meaning in the modern world, and disenchantment with established elites and structures, perceived or real, have motivated extremist movements and enabled them to recruit for a violent struggle against existing systems (See Box E: Mobility and the Fourth Industrial Revolution).

Hyper-connectivity does not naturally come together with greater tolerance or -adaptability, as seen in the reactions to the tragic human displacements that reached a historic high in 2015. However, the same hyper-connectivity also contains the potential to reach common ground based on greater acceptance and understanding of differences, which could help bring communities together rather than driving them apart. If we do not continue moving in this direction, however, the alternative is that it leads to increasing fragmentation.

Box E: Mobility and the Fourth Industrial Revolution

The movement of people around the world is both a significant phenomenon and a huge driver of wealth. How will the fourth industrial revolution impact human mobility? It may be too soon to tell, but extrapolating from current trends indicates that mobility will play an ever more important role in society and economics in the future than today:

– **Realizing life aspirations:** Corresponding to a rise in awareness of events and opportunities in other countries thanks to rising connectivity, mobility is increasingly seen as a life choice to be exercised at some point, especially by young people. While individual motivations vary enormously, the search for work, the desire to study, the need for protection, the desire to reunite family, and so on, there is a greater readiness to look for solutions over the horizon.

– **Redefining individual identities:** Individuals used to identify their lives most closely with a place, an ethnic group, a particular culture or even a language. The advent of online engagement and increased exposure to ideas from other cultures mean that identities are now more fungible than previously. People are now much more comfortable with carrying and managing multiple identities.

– **Redefining family identity:** Thanks to the combination of historical migration patterns and low-cost connectivity, family structures are being redefined. No longer bound by space, they often stretch across the world, with constant family dialogue, reinforced by digital means. Increasingly, the traditional family unit is being replaced by the trans-national family network.

– **Re-mapping labour markets:** Worker mobility has the potential to transform domestic labour markets for better or for worse. On one hand, workers in the developing world constitute a pool of human resources -- at multiple skill levels that can satisfy unmet labour market needs in the developed world. Talent mobility is a driver of creativity, of industrial innovation and work efficiency. On the other hand, the injection of migrant labour into domestic markets, if not managed effectively, can produce wage distortions and social unrest in host nations, while depriving origin countries of valuable human capital.

The digital revolution created new opportunities for communication and "mobility" that complemented and enhanced physical mobility. It is likely that the fourth industrial revolution will have a similar effect, as the fusion of the physical, digital and biological worlds will further transcend time/space limitations in such a way as to encourage mobility. One of the challenges of the fourth industrial revolution will therefore be the governance of human mobility to ensure that its benefits are fully realized by aligning sovereign rights and obligations with individual rights and aspirations, reconciling national and human security and finding ways to maintain social harmony in the midst of increasing diversity.

Source: Global Agenda Council on Migration, World Economic Forum

The changing nature of conflict

The fourth industrial revolution will affect the scale of conflict as well as its character. The distinctions between war and peace and who is a combatant and non-combatant are becoming uncomfortably blurred. Similarly, the battlefield is increasingly both local and global. Organizations such as Da'esh, or ISIS, operate principally in defined areas in the Middle East but they also recruit fighters from more than 100 countries, largely through social media, while related terrorist attacks can occur anywhere on the planet. Modern conflicts are increasingly hybrid in nature, combining traditional battlefield techniques with elements that were previously mostly associated with armed non-state actors. However, with technologies fusing in increasingly unpredictable ways and with state and armed non-state actors learning from each other, the potential magnitude of change is not yet widely appreciated.

As this process takes place and new, deadly technologies become easier to acquire and use, it is clear that the fourth industrial revolution offers individuals increasingly diverse ways to harm others on a grand scale. Realizing this leads to a greater sense of vulnerability.

It is not all bleak. Access to technology also brings with it the possibility of greater precision in warfare, cutting-edge protective wear for combat, the capacity to print essential spare parts or other components right on the battlefield, and so on.

Cyber warfare

Cyber warfare presents one of the most serious threats of our time. Cyberspace is becoming as much a theatre of engagement as land, sea and air was in the past. I can safely postulate that, while any future conflict between reasonably advanced actors may or may not play out in the physical world, it will most likely include a cyber-dimension simply because no modern opponent would resist the temptation to disrupt, confuse or destroy their enemy's sensors, communications and decision-making capability.

This will not only lower the threshold of war but will also blur the distinction between war and peace, because any networks or connected devices, from military systems to civilian infrastructure such as energy sources, electricity grids, health or traffic controls, or water supplies, can be hacked and attacked. The concept of the adversary is also affected as a result. Contrary to the past, you may not be certain of who is attacking you – and even whether you have been attacked at all. Defence, military and national security strategists focused on a limited number of traditionally hostile states, now they must consider a near-infinite and indistinct universe of hackers, terrorists, activists, criminals, and other possible foes. Cyber warfare can take many different forms – from criminal acts and espionage to destructive attacks such as Stuxnet – that remain largely underestimated and misunderstood because they are so new and difficult to counter.

Since 2008, there have been many instances of cyber attacks directed at both specific countries and companies, yet discussions about this new era of warfare are still in their infancy and the gap between those who understand the highly technical issues of cyber warfare and those who are developing cyber policy widens by the day. Whether a set of shared norms will evolve for cyber warfare, analogous to those developed for nuclear, biological and chemical weapons, remains an open question. We lack even a taxonomy to agree on what amounts to an attack and the appropriate response, with what and by whom. Part of the equation to manage this scenario is to define what data travels across borders. This is an indication of how far there is to go on effectively controlling cross-border cyber based transactions without inhibiting the positive outputs from a more interconnected world.

Autonomous warfare

Autonomous warfare, including the deployment of military robots and AI-powered automated weaponry, creates the prospect of "robo-war", which will play a transformative role in future conflict.

The seabed and space are also likely to become increasingly militarized, as more and more actors – state and commercial – gain the ability to send up satellites and mobilize unmanned underwater vehicles capable of disrupting fibre-optic cables and satellite traffic. Criminal gangs are already using off-the-shelf quadrocopter drones to spy on and attack rivals. Autonomous weapons, capable of identifying targets and deciding to open fire without human intervention, will become increasingly feasible, challenging the laws of war.

Box F: Emerging Technologies Transforming International Security

Drones: They are essentially flying robots. The US currently leads but the technology is spreading widely and becoming more affordable.

Autonomous weapons: Combining drone technology with artificial intelligence, they have the potential to select and engage targets without human intervention, according to pre-defined criteria.

Militarization of space: While more than half of all satellites are commercial, these orbiting communications devices are increasingly important for military purposes. A new generation of hypersonic "glide" weapons are also poised to enter this domain, increasing the probability that space will play a role in future conflicts and raising concern that current mechanisms to regulate space activities are no longer sufficient.

Wearable devices: They can optimize health and performance under conditions of extreme stress or produce exoskeletons that enhance soldiers' performance, allowing a human to carry loads of around 90 kg without difficulty.

Additive manufacturing: It will revolutionize supply chains by enabling replacement parts to be manufactured in the field from digitally transmitted designs and locally available materials. It could also enable the development of new kinds of warheads, with greater control of particle size and detonation.

Renewable energy: This enables power to be generated locally, revolutionizing supply chains and enhancing the capacity to print parts on demand in even remote locations.

Nanotechnology: Nano is progressively leading to metamaterials, smart materials which possess properties that do not occur naturally. It will make weaponry better, lighter, more mobile, smarter and more precise, and will ultimately result in systems that can self-replicate and assemble.

Biological weapons: The history of biological warfare is nearly as old as the history of warfare itself, but rapid advances in biotechnology, genetics and genomics are the harbinger of new highly lethal weapons. Airborne designer viruses, engineered superbugs, genetically modified plagues and so on: all these form the basis of potential doomsday scenarios.

Biochemical weapons: As with biological weapons, technological innovation is making the assembly of these weapons almost as easy as a do-it-yourself task. Drones could be employed to deliver them.

Social Media: While digital channels provide opportunities for spreading information and organizing action for good causes, they can also be used to spread malicious content and propaganda and, as with ISIS, employed by extremist groups to recruit and mobilize followers. Young adults are particularly vulnerable, especially if they lack a stable social support network.

Many of the technologies described in Box F: Emerging Technologies Transforming International Security already exist. As an example, Samsung's SGR-A1 robots, equipped with two machine guns and a gun with rubber bullets, now man border posts in the Korean Demilitarized Zone. They are, for the moment, controlled by human operators but could, once programmed, identify and engage human targets independently.

Last year, the UK Ministry of Defence and BAE Systems announced the successful test of the Taranis stealth plane, known also as Raptor, which can take off, fly to a given destination and find a set target with little intervention from its operator unless required. There are many such examples.[50] They will multiply, and in the process, raise critical questions at the intersection of geopolitics, military strategy and tactics, regulation and ethics.

New frontiers in global security

As stressed several times in this book, we only have a limited sense of the ultimate potential of new technologies and what lies ahead. This is no less the case in the realm of international and domestic security. For each innovation we can think of, there will be a positive application and a possible dark side. While neurotechnologies such as neuroprosthetics are already employed to solve medical problems, in future they could be applied to military purposes. Computer systems attached to brain tissue could enable a paralysed patient to control a robotic arm or leg. The same technology could be used to direct a bionic pilot or soldier. Brain devices designed to treat the conditions of Alzheimer's disease could be implanted in soldiers to erase memories or create new ones. "It's not a question of if non-state actors will use some form of neuroscientific techniques or technologies, but when, and which ones they'll use," reckons James Giordano, a neuroethicist at Georgetown University Medical Center, "The brain is the next battlespace."[51]

The availability and, at times, the unregulated nature of many of these innovations have a further important implication. Current trends suggest a rapid and massive democratization of the capacity to inflict damage on a very large scale, something previously limited to governments and very sophisticated organizations. From 3D-printed weapons to genetic engineering in home laboratories, destructive tools across a range of emerging technologies are becoming more readily available. And with the fusion of technologies, a key theme of this book, unpredictable dynamics inherently surface, challenging existing legal and ethical frameworks.

Towards a more secure world

In the face of these challenges, how do we persuade people to take the security threats from emerging technologies seriously? Even more importantly, can we engender cooperation between the public and private sectors on the global scale to mitigate these threats?

Over the second half of the last century, the fear of nuclear warfare gradually gave way to the relative stability of mutually assured destruction (MAD), and a nuclear taboo seems to have emerged.

If the logic of MAD has worked so far it is because only a limited number of entities possessed the power to destroy each other completely and they balanced each other out. A proliferation of potentially lethal actors, however, could undermine this equilibrium, which was why nuclear states agreed to cooperate to keep the nuclear club small, negotiating the Treaty on the Non-Proliferation of Nuclear Weapons (NPT) in the late 1960s.

While they disagreed on most other issues, the Soviet Union and the United States understood that their best protection laid in remaining vulnerable to each other. This led to the Anti-Ballistic Missile Treaty (ABMT), effectively limiting the right to take defensive measures against missile-delivered nuclear weapons. When destructive capacity is no longer limited to a handful of entities with broadly similar resources, tactics and interests in preventing escalation doctrines such as MAD are less relevant.

Driven by the changes heralded by the fourth industrial revolution, could we discover some alternative equilibrium that analogously turns vulnerability into stability and security? Actors with very different perspectives and interests need to be able to find some kind of *modus vivendi* and cooperate in order to avoid negative proliferation.

Concerned stakeholders must cooperate to create legally binding frameworks as well as self-imposed peer-based norms, ethical standards and mechanisms to control potentially damaging emerging technologies, preferably without impeding the capacity of research to deliver innovation and economic growth.

International treaties will surely be needed, but I am concerned that regulators in this field will find themselves running behind technological advances, due to their speed and multifaceted impact. Hence, conversations among educators and developers about the ethical standards that should apply to emerging technologies of the fourth industrial revolution are urgently needed to establish common ethical guidelines and embed them in society and culture. With governments and government based structures, lagging behind in the regulatory space, it may actually be up to the private sector and non-state actors to take the lead.

The development of new warfare technologies is, understandably, taking place in a relatively isolated sphere. One concern I have, however, is the potential retreat of other sectors, such as gene-based medicine and research, into isolated, highly-specialized spheres, thereby lowering our collective ability to discuss, understand and manage both challenges and opportunities.

3.4 Society

Scientific advancement, commercialization and the diffusion of innovation are social processes that unfold as people develop and exchange ideas, values, interests and social norms in a variety of contexts. This makes it hard to discern the full societal impact of new technological systems: there are many intertwined components that comprise our societies and many innovations that are in some way co-produced by them.

The big challenge for most societies will be how to absorb and accommodate the new modernity while still embracing the nourishing aspects of our traditional value systems. The fourth industrial revolution, which tests so many of our fundamental assumptions, may exacerbate the tensions which exist between deeply religious societies defending their fundamental values and those whose beliefs are shaped by a more secular worldview. The greatest danger to global cooperation and stability may come from radical groups fighting progress with extreme, ideologically motivated violence.

As sociologist Manuel Castells, professor of communication technology and society at the Annenberg School of Communication and Journalism at the University of Southern California, has noted: "In all moments of major technological change, people, companies, and institutions feel the depth of the change, but they are often overwhelmed by it, out of sheer ignorance of its effects".[52] Being overwhelmed due to ignorance is precisely what we should avoid, particularly when it comes to how the many diverse communities that comprise modern society form, develop and relate to one another.

The previous discussion about the different impacts of the fourth industrial revolution on the economy, business, geopolitics and international security, regions and cities makes it clear that the new technological revolution will have multiple influences on society. In the next section, I will explore two of the most important drivers of change – how the potential for rising inequality puts pressure on the middle class, and how the integration of digital media is changing how communities form and relate to one another.

3.4.1 Inequality and the middle class

The discussion on economic and business impacts highlighted a number of different structural shifts which have contributed to rising inequality to date, and which may be further exacerbated as the fourth industrial revolution unfolds. Robots and algorithms increasingly substitute capital for labour, while investing (or more precisely, building a business in the digital economy) becomes less capital intensive. Labour markets, meanwhile, are becoming biased towards a limited range of technical skill sets, and globally connected digital platforms and marketplaces are granting outsized rewards to a small number of "stars". As all these trends happen, the winners will be those who are able to participate fully in innovation-driven ecosystems by providing new ideas, business models, products and services, rather than those who can offer only low-skilled labour or ordinary capital.

These dynamics are why technology is regarded as one of the main reasons incomes have stagnated, or even decreased, for a majority of the population in high-income countries. Today, the world is very unequal indeed. According to Credit Suisse's Global Wealth Report 2015, half of all assets around the world are now controlled by the richest 1% of the global population, while "the lower half of the global population collectively own less than 1% of global wealth".[53] The Organisation for Economic Co-operation and Development (OECD) reports that the average income of the richest 10% of the population in OECD countries is approximately nine times that of the poorest 10%.[54] Further, inequality within most countries is rising, even in those that have experienced rapid growth across all income groups and dramatic drops in the number of people living in poverty. China's Gini Index, for example, rose from approximately 30 in the 1980s to over 45 by 2010.[55]

Rising inequality is more than an economic phenomenon of some concern – it is a major challenge for societies. In their book *The Spirit Level: Why Greater Equality Makes Societies Stronger*, British epidemiologists Richard Wilkinson and Kate Pickett put forward data indicating that unequal societies tend to be more violent, have higher numbers of people in prison, experience greater levels of mental illness and obesity, and have lower life expectancies and lower levels of trust. The corollary, they found, is that, after controlling for average incomes, more equal societies have higher levels of child well-being, lower levels of stress and drug use, and lower infant mortality.[56] Other researchers have found that higher levels of inequality increase segregation and reduce educational outcomes for children and young adults.[57]

While the empirical data are less certain, there are also widespread fears that higher levels of inequality lead to higher levels of social unrest. Among the 29 global risks and 13 global trends identified in the Forum's *Global Risks Report 2016*, the strongest interconnections occur between rising income disparity, unemployment or underemployment and profound social instability. As discussed further below, a world of greater connectivity and higher expectations can create significant social risks if populations feel they have no chance of attaining any level of prosperity or meaning in their lives.

Today, a middle-class job no longer guarantees a middle-class lifestyle, and over the past 20 years, the four traditional attributes of middle-class status (education, health, pensions and house ownership) have performed worse than inflation. In the US and the UK, education is now priced as a luxury. A winner-takes-all market economy, to which the middle-class has increasingly limited access, may percolate into democratic malaise and dereliction which compound social challenges.

3.4.2 Community

From a broad societal standpoint, one of the greatest (and most observable) effects of digitization is the emergence of the "me-centred" society – a process of individuation and emergence of new forms of belonging and community. Contrary to the past, the notion of belonging to a community today is more defined by personal projects and individual values and interests rather than by space (the local community), work and family.

New forms of digital media, which form a core component of the fourth industrial revolution, are increasingly driving our individual and collective framing of society and community. As the Forum explores in its *Digital Media and Society* report, digital media is connecting people one-to-one and one-to-many in entirely new ways, enabling users to maintain friendships across time and distance, creating new interest groups and enabling those who are socially or physically isolated to connect with like-minded people. The high availability, low costs and geographically neutral aspects of digital media also enable greater interaction across social, economic, cultural, political, religious and ideological boundaries.

Access to online digital media creates substantial benefits for many. Beyond its role in providing information (for example, refugees fleeing Syria use Google Maps and Facebook groups not only to plan travel routes but also to avoid being exploited by human traffickers[58]), it also provides opportunities for individuals to have a voice and participate in civic debate and decision-making.

Unfortunately, while the fourth industrial revolution empowers citizens, it can also be used to act against their interests. The Forum's *Global Risks Report 2016* describes the phenomenon of the "(dis)empowered citizen", whereby individuals and communities are simultaneously empowered and excluded by the use of emerging technologies by governments, companies and interest groups (see Box G: The (Dis)empowered Citizen).

The democratic power of digital media means it can also be used by non-state actors, particularly communities with harmful intentions to spread propaganda and to mobilize followers in favour of extremist causes, as has been seen recently with the rise of Da'esh and other social-media-savvy terrorist organizations.

There is the danger that the dynamics of sharing that typifies social media use can skew decision-making and pose risks to civil society. Counter-intuitively, the fact that there is so much media available through digital channels can mean that an individual's news sources become narrowed and polarised into what MIT clinical psychologist Sherry Turkle, a professor of the social studies of science and technology, calls a "spiral of silence". This matters because what we read, share and see in the context of social media shapes our political and civic decisions.

Box G: The (Dis)empowered Citizen

The term "(dis)empowered citizen" describes the dynamic emerging from the interplay of two trends: one empowering, one disempowering. Individuals feel empowered by changes in technology that make it easier for them to gather information, communicate and organize, and are experiencing new ways to participate in civic life. At the same time, individuals, civil society groups, social movements and local communities feel increasingly excluded from meaningful participation in traditional decision-making processes, including voting and elections, and disempowered in terms of their ability to influence and be heard by the dominant institutions and sources of power in national and regional governance.

At its most extreme, there is the very real danger that governments might employ combinations of technologies to suppress or oppress actions of civil society organizations and groups of individuals who seek to create transparency around the activities of governments and businesses and promote change. In many countries around the world there is evidence that the space for civil society is shrinking as governments promote legislation and other policies which restrict the independence of civil society groups and restrict their activities. The tools of the fourth industrial revolution enable new forms of surveillance and other means of control that run counter to healthy, open societies.

Source: *Global Risks Report 2016,* World Economic Forum

As an example, a study of the impact of get-out-the-vote messages on Facebook found that they "increased turnout directly by about 60,000 voters and indirectly through social contagion by another 280,000 voters, for a total of 340,000 additional votes."[59] This research highlights the power that digital media platforms have in selecting and promoting the media we consume online. It also indicates the opportunity for online technologies to blend traditional forms of civic engagement (such as voting for local, regional or national representatives) with innovative ways to give citizens more direct influence over decisions that affect their communities.

As with almost all the impacts addressed in this section, it is clear that the fourth industrial revolution brings great opportunities while also posing significant risks. One of the key tasks the world faces as this revolution emerges is how to gather more and better data on both the benefits and challenges to community cohesion.

3.5 The Individual

The fourth industrial revolution is not only changing what we do but also who we are. The impact it will have on us as individuals is manifold, affecting our identity and its many related facets – our sense of privacy, our notions of ownership, our consumption patterns, the time we devote to work and leisure, how we develop our careers, cultivate our skills. It will influence how we meet people and nurture relationships, the hierarchies upon which we depend, our health, and maybe sooner than we think, it could lead to forms of human augmentation that cause us to question the very nature of human existence. Such changes elicit excitement and fear as we move at unprecedented speed.

Until now, technology has primarily enabled us to do things in easier, faster and more efficient ways. It has also provided us with opportunities for personal development. But we are beginning to see that there is much more on offer and at stake. For all the reasons already mentioned, we are at the threshold of a radical systemic change that requires human beings to adapt continuously. As a result, we may witness an increasing degree of polarization in the world, marked by those who embrace change versus those who resist it.

This gives rise to an inequality that goes beyond the societal one described earlier. This ontological inequality will separate those who adapt from those who resist – the material winners and losers in all senses of the word. The winners may even benefit from some form of radical human improvement generated by certain segments of the fourth industrial revolution (such as genetic engineering) from which the losers will be deprived. This risks creating class conflicts and other clashes unlike anything we have seen before. This potential division and the tensions it stirs will be exacerbated by a generational divide caused by those who have only known and grown up in a digital world versus those who have not and who must adapt. It also gives rise to many ethical issues.

As an engineer, I am a great technology enthusiast and early adopter. Yet I wonder, as many psychologists and social scientists do, how the inexorable integration of technology in our lives will impact our notion of identity and whether it could diminish some of our quintessential human capacities such as self-reflection, empathy and compassion.

3.5.1 Identity, Morality and Ethics

The mind-boggling innovations triggered by the fourth industrial revolution, from biotechnology to AI, are redefining what it means to be human. They are pushing the current thresholds of lifespan, health, cognition and capabilities in ways that were previously the preserve of science fiction. As knowledge and discoveries in these fields progress, our focus and commitment to having ongoing moral and ethical discussions is critical. As human beings and as social animals, we will have to think individually and collectively about how we respond to issues such as life extension, designer babies, memory extraction and many more.

At the same time, we must also realize that these incredible discoveries could also be manipulated to serve special interests – and not necessarily those of the public at large. As theoretical physicist and author Stephen Hawking and fellow scientists Stuart Russell, Max Tegmark and Frank Wilczek wrote in the newspaper *The Independent* when considering the implications of artificial intelligence: "Whereas the short-term impact of AI depends on who controls it, the long-term impact depends on whether it can be controlled at all…All of us should ask ourselves what we can do now to improve the chances of reaping the benefits and avoiding the risks".[60]

One interesting development in this area is OpenAI, a non-profit AI research company announced in December 2015 with the goal to "advance digital intelligence in the way that is most likely to benefit humanity as a whole, unconstrained by a need to generate financial return".[61] The initiative – chaired by Sam Altman, President of Y Combinator, and Elon Musk, CEO of Tesla Motors - has secured $1 billion in committed funding. This initiative underscores a key point made earlier – namely, that one of the biggest impacts of the fourth industrial revolution is the empowering potential catalyzed by a fusion of new technologies. Here, as Sam Altman stated, "the best way AI can develop is if it's about individual empowerment and making humans better, and made freely available to everyone."[62]

The human impact of some particular technologies such as the internet or smart phones is relatively well understood and widely debated among experts and academics. Other impacts are so much harder to grasp. Such is the case with AI or synthetic biology. We may see designer babies in the near future, along with a whole series of other edits to our humanity – from eradicating genetic diseases to augmenting human cognition. These will raise some of the biggest ethical and spiritual questions we face as human beings (see Box H: On the Ethical Edge).

Box H: On the Ethical Edge

Technological advances are pushing us to new frontiers of ethics. Should we use the staggering advances in biology only to cure disease and repair injury, or should we also make ourselves better humans? If we accept the latter, we risk turning parenthood into an extension of the consumer society, in which case could our children become commoditized as made-to-order objects of our desire? And what does it mean to be "better"? To be disease free? To live longer? To be smarter? To run faster? To have a certain appearance?

We face similarly complex and on-the-edge questions with artificial intelligence. Consider the possibility of machines thinking ahead of us or even out-thinking us. Amazon and Netflix already possess algorithms that predict which films and books we may wish to watch and read. Dating and job placement sites suggest partners and jobs – in our neighbourhood or anywhere in the world – that their systems figure might suit us best. What do we do? Trust the advice provided by an algorithm or that offered by family, friends or colleagues? Would we consult an AI-driven robot doctor with a perfect or near-perfect diagnosis success rate – or stick with the human physician with the assuring bedside manner who has known us for years?

When we consider these examples and their implications for humans, we are in uncharted territory – the dawn of a human transformation unlike anything we have experienced before.

Another substantial issue relates to the predictive power of artificial intelligence and machine learning. If our own behaviour in any situation becomes predictable, how much personal freedom would we have or feel that we have to deviate from the prediction? Could this development potentially lead to a situation where human beings themselves begin to act as robots? This also leads to a more philosophical question: How do we maintain our individuality, the source of our diversity and democracy, in the digital age?

3.5.2 Human Connection

As the ethical questions raised above suggest, the more digital and high-tech the world becomes, the greater the need to still feel the human touch, nurtured by close relationships and social connections. There are growing concerns that, as the fourth industrial revolution deepens our individual and collective relationships with technology, it may negatively affect our social skills and ability to empathize. We see this already happening. A 2010 study by a research team at the University of Michigan found a 40%

decline in empathy among college students (as compared to their counterparts 20 or 30 years ago), with most of this decline coming after 2000.[63]

According to MIT's Sherry Turkle, 44% of teenagers never unplug, even while playing sports or having a meal with family or friends. With face-to-face conversations crowded out by online interactions, there are fears that an entire generation of young people consumed by social media is struggling to listen, make eye contact or read body language.[64]

Our relationship with mobile technologies is a case in point. The fact that we are always connected may deprive us of one of our most important assets: the time to pause, reflect and engage in a substantive conversation neither aided by technology nor intermediated by social media. Turkle refers to studies showing that, when two people are talking, the mere presence of a phone on the table between them or in their peripheral vision changes both what they talk about and their degree of connectedness.[65] This does not mean we give up our phones but rather that we use them "with greater intention".

Other experts express related concerns. Technology and culture writer Nicholas Carr states that the more time we spend immersed in digital waters, the shallower our cognitive capabilities become due to the fact that we cease exercising control over our attention: "The Net is by design an interruption system, a machine geared for dividing attention. Frequent interruptions scatter our thoughts, weaken our memory, and make us tense and anxious. The more complex the train of thought we're involved in, the greater the impairment the distractions cause."[66]

Back in 1971, Herbert Simon, who won the Nobel Prize in Economics in 1978, warned that "a wealth of information creates a poverty of attention." This is much worse today, in particular for decision-makers who tend to be overloaded with too much "stuff" – overwhelmed and on overdrive, in a state of constant stress. "In an age of acceleration, nothing can be more exhilarating than going slow," writes the travel essayist Pico Iyer. "And in an age of distraction, nothing is so luxurious as paying attention. And in an age of constant movement, nothing is so urgent as sitting still."[67]

Our brain, engaged by all the digital instruments that connect us on a 24-hour basis, risks becoming a perpetual-motion machine that puts in an unremitting frenzy. It is not unusual for me to talk to leaders who say that they no longer have time to pause and reflect, let alone enjoy the "luxury" of reading even a short article all the way through. Decision-makers from all parts of global society seem to be in a state of ever-increasing exhaustion, so deluged by multiple competing demands that they turn from frustration to resignation and sometimes despair. In our new digital age, it is indeed difficult to step back, though not impossible.

3.5.3 Managing Public and Private Information

One of the greatest individual challenges posed by the internet, and our increasing degree of interconnectedness in general, concerns privacy. It is an issue that looms larger and larger because, as the Harvard University political philosopher Michael Sandel has observed "we seem to be increasingly willing to trade privacy for convenience with many of the devices that we routinely use".[68] Spurred in part by the revelations of Edward Snowden, the global debate about the meaning of privacy in a world of greater transparency has only just begun, as we see how the internet can be an unprecedented tool of liberation and democratization and at the same time, an enabler of indiscriminate, far-reaching and almost unfathomable mass surveillance.

Why does privacy matter so much? We all instinctively understand why privacy is so essential for our individual selves. Even for those who claim that they do not particularly value privacy and have nothing to hide, there are all sorts of things said and done that we may not want anyone else to know about. There is abundant research showing that when someone knows he is being watched, his behaviour becomes more conformist and compliant.

This book, however, is not the place to engage in a lengthy reflection about the meaning of privacy or to respond to questions about data ownership. I fully expect, however, that a debate about many fundamental issues such as the impact on our inner lives, stemming from the loss of control over our data, will only intensify in the years ahead (see Box I: Wellness and the Bounds of Privacy).

These issues are incredibly complex. We are just starting to get a sense of their possible psychological, moral and social implications. On a personal level, I foresee the following problem related to privacy: When one's life becomes fully transparent and when indiscretions big or small become knowable to all, who will have the courage to assume top leadership responsibilities?

The fourth industrial revolution renders technology an all-pervasive and predominant part of our individual lives, and yet we are only just starting to understand how this technological sea-change will affect our inner selves. Ultimately, it is incumbent upon each of us to guarantee we are served, not enslaved, by technology. At a collective level, we must also ensure that the challenges technology throws at us are properly understood and analysed. Only in this way can we be certain that the fourth industrial revolution will enhance, rather than damage, our wellbeing.

Box I: Wellness and the Bounds of Privacy

What is currently happening with wearable wellness devices provides a sense of the complexity of the privacy issue. An increasing number of insurance companies are considering making this offer to their policyholders: If you wear a device that monitors your wellness – how much you sleep and exercise, the number of steps you take each day, the number and type of calories you eat, etc. – and if you agree that this information can be sent to your health insurance provider, we will offer you a discount on your premium.

Is this a development we should welcome because it motivates us to live healthier lives? Or is it a worrisome move towards a way of life where surveillance – from government and companies alike – becomes ever more intrusive? For the moment, this example refers to an individual choice – the decision to accept wearing a wellness device or not.

But pushing this further, let us assume that it is now the employer that directs each of its staff to wear a device that reports health data to the insurer because the company wants to improve productivity and possibly to decrease its health insurance costs. What if the company requires reluctant employees to abide or else pay a fine? What previously seemed like a conscious individual choice – wearing a device or not – becomes a matter of conforming to new social norms that one may deem unacceptable.

The Way Forward

The fourth industrial revolution may be driving disruption, but the challenges it presents are of our own making. It is thus in our power to address them and enact the changes and policies needed to adapt (and flourish) in our emerging new environment.

We can only meaningfully address these challenges if we mobilize the collective wisdom of our minds, hearts and souls. To do so, I believe we must adapt, shape and harness the potential of disruption by nurturing and applying four different types of intelligence:

- contextual (the mind) – how we understand and apply our knowledge
- emotional (the heart) – how we process and integrate our thoughts and feelings and relate to ourselves and to one another
- inspired (the soul) – how we use a sense of individual and shared purpose, trust, and other virtues to effect change and act towards the common good
- physical (the body) – how we cultivate and maintain our personal health and well-being and that of those around us to be in a position to apply the energy required for both individual and systems transformation

Contextual intelligence – the mind

Good leaders understand and master contextual intelligence.[69] A sense of context is defined as the ability and willingness to anticipate emerging trends and connect the dots. These have been common characteristics of effective leadership across generations and, in the fourth industrial revolution, they are a prerequisite for adaptation and survival.

To develop contextual intelligence, decision-makers must first understand the value of diverse networks. They can only confront significant levels of disruption if they are highly connected and well networked across traditional boundaries. Decision-makers must possess a capacity and readiness to engage with all those who have a stake in the issue at hand. In this way, we should aspire to be more connected and inclusive.

It is only by bringing together and working in collaboration with leaders from business, government, civil society, faith, academia and the young generation that it becomes possible to obtain a holistic perspective of what is going on. In addition, this is critical to develop and implement integrated ideas and solutions that will result in sustainable change.

This is the principle embedded in the multistakeholder theory (what the World Economic Forum communities often call the Spirit of Davos), which I first proposed in a book published in 1971.[70] Boundaries between sectors and professions are artificial and are proving to be increasingly counterproductive. More than ever, it is essential to dissolve these barriers by engaging the power of networks to forge effective partnerships. Companies and organizations that fail to do this and do not walk the talk by building diverse teams will have a difficult time adjusting to the disruptions of the digital age.

Leaders must also prove capable of changing their mental and conceptual frameworks and their organising principles. In today's disruptive, fast-changing world, thinking in silos and having a fixed view of the future is fossilizing, which is why it is better, in the dichotomy presented by the philosopher Isaiah Berlin in his 1953 essay about writers and thinkers, to be a fox than a hedgehog. Operating in an increasingly complex and disruptive environment requires the intellectual and social agility of the fox rather than fixed and narrow focus of the hedgehog. In practical terms, this means that leaders cannot afford to think in silos. Their approach to problems, issues and challenges must be holistic, flexible and adaptive, continuously integrating many diverse interests and opinions.

Emotional intelligence – the heart

As a complement to, not a substitute for, contextual intelligence, emotional intelligence is an increasingly essential attribute in the fourth industrial revolution. As management psychologist David Caruso of the Yale Center for Emotional Intelligence has stated, it should not be seen as the opposite of rational intelligence or "the triumph of heart over head – it is the unique intersection of both."[71] In academic literature, emotional intelligence is credited with allowing leaders to be more innovative and enabling them to be agents of change.

For business leaders and policymakers, emotional intelligence is the vital foundation for skills critical to succeed in the era of the fourth industrial revolution, namely self-awareness, self-regulation, motivation, empathy and social skills.[72] Academics who specialize in the study of emotional intelligence show that great decision-makers are differentiated from average ones by their level of emotional intelligence and capacity to cultivate this quality continuously.

In a world characterized by persistent and intense change, institutions rich in leaders with high emotional intelligence will not only be more creative but will also be better equipped to be more agile and resilient – an essential trait for coping with disruption. The digital mindset, capable of institutionalizing cross-functional collaboration, flattening hierarchies, and building environments that encourage a generation of new ideas is profoundly dependent on emotional intelligence.

Inspired intelligence – the soul

Alongside contextual and emotional intelligence, there is a third critical component for effectively navigating the fourth industrial revolution. It is what I call inspired intelligence. Drawing from the Latin *spirare,* to breathe, inspired intelligence is about the continuous search for meaning and purpose. It focuses on nourishing the creative impulse and lifting humanity to a new collective and moral consciousness based on a shared sense of destiny.

Sharing is the key idea here. As I mentioned previously, if technology is one of the possible reasons why we are moving towards a me-centred society, it is an absolute necessity that we rebalance this trend towards a focus on the self with a pervasive sense of common purpose. We are all in this together and risk being unable to tackle the challenges of the fourth industrial revolution and reap the full benefits of the fourth industrial revolution unless we collectively develop a sense of shared purpose.

To do this, trust is essential. A high level of trust favours engagement and teamwork, and this is made all the more acute in the fourth industrial revolution, where collaborative innovation is at the core. This process can only take place if it is nurtured in an environment of trust because there are so many different constituents and issues involved. Ultimately, all stakeholders have a role in ensuring that innovation is directed to the common good. If any major group of stakeholders feels that this is not the case, trust will be eroded.

In a world where nothing is constant anymore, trust becomes one of the most valuable attributes. Trust can only be earned and maintained if decision makers are embedded within a community, and taking decisions always in the common interest and not in pursuit of individual objectives.

Physical intelligence – the body

Contextual, emotional and inspired intelligence are all essential attributes for coping with, and benefitting from, the fourth industrial revolution. They will, however, require the vital support of a fourth form of intelligence – the physical one, which involves supporting and nourishing personal health and well-being. This is critical because as the pace of change accelerates, as complexity increases, and as the number of players involved in our decision-making processes increases, the need to keep fit and remain calm under pressure becomes all the more essential.

Epigenetics, a field of biology that has flourished in recent years, is the process through which the environment modifies the expression of our genes. It shows incontrovertibly the critical importance of sleep, nutrition and exercise in our lives. Regular exercise, for example, has a positive impact on the way we think and feel. It directly affects our performance at work and ultimately, our ability to succeed.

Understanding and grasping new ways of keeping our physical bodies in harmony with our mind, our emotions, and the world at-large is incredibly important, and we are learning more about this through the incredible advances being made in numerous areas, including medical sciences, wearable devices, implantable technologies and brain research. In addition, I often say that a leader requires "good nerves" to address effectively the many simultaneous and complex challenges that we are facing. This will be increasingly critical in order to navigate and harness the opportunities of the fourth industrial revolution.

Towards a new cultural renaissance

As the poet Rainer Maria Rilke wrote, "the future enters into us...in order to transform itself in us long before it happens."[73] We must not forget that the era we currently live in, the Anthropocene or Human Age, marks the first time in the history of the world that human activities are the primary force in shaping all life-sustaining systems on earth.

It is up to us.

Today we find ourselves at the beginning of the fourth industrial revolution, looking forward and, more importantly, possessing the ability to influence its path.

Knowing what is required to thrive is one thing; acting upon it is another. Where is all this leading and how can we best prepare?

Voltaire, the French philosopher and writer of the Enlightenment era who lived for many years just a few miles away from where I am writing this book, once said: "Doubt is an uncomfortable condition, but certainty is a ridiculous one."[74] Indeed, it would be naive to claim that we know exactly where the fourth industrial revolution will lead. But it would be equally naive to be paralysed by fear and uncertainty about what that direction might be. As I have emphasized throughout this book, the eventual course that the fourth industrial revolution takes will ultimately be determined by our ability to shape it in a way that unleashes its full potential.

Clearly, the challenges are as daunting as the opportunities are compelling. Together, we must work to transform these challenges into opportunities by adequately – and proactively – preparing for their effects and impact. The world is fast changing, hyper-connected, ever more complex and becoming more fragmented but we can still shape our future in a way that benefits all. The window of opportunity for doing so is now.

As a first and vital step, we must continue to raise awareness and drive understanding across all sectors of society, which is what this book aspires to achieve. We must stop thinking in compartmentalized ways when making decisions – particularly as the challenges we face are increasingly interconnected. Only an inclusive approach can engender the understanding required to address the many issues raised by the fourth industrial revolution. This will require collaborative and flexible structures that reflect the integration of various ecosystems and which take fully into account all stakeholders, bringing together the public and private sectors, as well as the most knowledgeable minds in the world from all backgrounds.

Second, building on a shared understanding, we need to develop positive, common and comprehensive narratives about how we can shape the fourth industrial revolution for current and future generations. Although we may not know the precise content of these narratives, we do know critical features that they must contain. For example, they must make explicit the values and ethical principles that our future systems must embody. Markets are effective drivers of wealth creation, but we must ensure that values and ethics are at the heart of our individual and collective behaviours, and the systems they nourish. These narratives must also evolve progressively higher degrees of perspective-taking, from tolerance and respect to care and compassion. They should also be empowering and inclusive, driven by shared values that encourage this.

Third, on the basis of raised awareness and shared narratives, we must embark on restructuring our economic, social and political systems to take full advantage of the opportunities presented. It is clear that our current decision-making systems and dominant models of wealth creation were designed and incrementally evolved throughout the first three industrial revolutions. These systems, however, are no longer equipped to deliver on the current, and more to the point, the future generational needs in the context of the fourth industrial revolution. This will clearly require systemic innovation and not small-scale adjustments or reforms at the margin.

As all three steps show, we cannot get there without ongoing cooperation and dialogue - at local, national and supra-national levels, with all interested parties having a voice. We need to focus on getting the underlying conditions right, and not just concentrate on the technical aspects. As the evolutionist Martin Nowak, a professor of mathematics and biology at Harvard University, reminds us, cooperation is "the only thing that will redeem mankind."[75] As the principal architect of four billion years of evolution, cooperation has been a driving force because it enables us to adapt amid increasing complexity and strengthens political, economic and social cohesion through which substantial progress is achieved.

With effective multistakeholder cooperation, I am convinced that the fourth industrial revolution has the potential to address – and possibly solve – the major challenges that the world currently faces.

In the end, it comes down to people, culture and values. Indeed, we need to work very hard to ensure that all citizens across cultures, nations and income groups understand the need to master the fourth industrial revolution and its civilizational challenges.

Let us together shape a future that works for all by putting people first, empowering them and constantly reminding ourselves that all of these new technologies are first and foremost tools made by people for people.

Let us therefore take collective responsibility for a future where innovation and technology are centred on humanity and the need to serve the public interest, and ensure that we employ them to drive us all towards more sustainable development.

We can go even further. I firmly believe that the new technology age, if shaped in a responsive and responsible way, could catalyse a new cultural renaissance that will enable us to feel part of something much larger than ourselves – a true global civilization. The fourth industrial revolution has the potential to robotize humanity, and thus compromise our traditional sources of meaning - work, community, family, identity. Or we can use the fourth industrial revolution to lift humanity into a new collective and moral consciousness based on a shared sense of destiny. It is incumbent on us all to make sure that the latter is what happens.

Acknowledgements

All of us at the World Economic Forum are aware of our responsibility, as the international organization for public private cooperation, to serve as a global platform to help define the challenges associated the fourth industrial revolution and help all stakeholders shape appropriate solutions in a proactive and comprehensive manner, in collaboration with our partners, members, and constituents.

For this reason, the theme of the Forum's Annual Meeting 2016 in Davos-Klosters is "Mastering the Fourth Industrial Revolution". We are committed to catalysing constructive discussions and partnerships around this topic across all our challenges, projects and meetings. The Forum's Annual Meeting of New Champions in Tianjin, China, in June 2016, will also provide a critical opportunity for leaders and innovators across research, technology, commercialization and regulation to meet and exchange ideas about how to harness the fourth industrial revolution to the greatest possible benefit of all. For all these activities, I hope this book serves as a primer and guide, equipping leaders to grapple with the political, social and economic implications as well as to understand the advances in technology that create them.

This book would not have been possible without the enthusiastic support and engagement of all my colleagues at the World Economic Forum. I owe them immense thanks. I express my particular gratitude to Nicholas Davis, Thierry Malleret and Mel Rogers who were essential partners throughout the research and writing process. I am also thankful to my colleagues and all the teams who contributed to specific sections of the book, particularly Jennifer Blanke, Margareta Drzeniek-Hanouz, Silvia Magnoni and Saadia Zahidi on economics and society; Jim Hagemann Snabe, Mark Spelman and Bruce Weinelt on business and industry; Dominic Waughray on the environment; Helena Leurent on governments; Espen Barth Eide and Anja Kaspersen on geopolitics and international security; and Olivier Oullier on neurotechnology.

Writing this book uncovered exceptional expertise across the whole Forum staff, and I thank everyone who shared their ideas with me, both online and in person. Here, in particular, I thank members of the Emerging Technologies taskforce: David Gleicher, Rigas Hadzilacos, Natalie Hatour, Fulvia Montresor and Olivier Woeffray – and the many others who spent time thinking deeply about these issues: Chidiogo Akunyili, Claudio Cocorocchia, Nico Daswani, Mehran Gul, Alejandra Guzman, Mike Hanley, Lee Howell, Jeremy Jurgens, Bernice Lee, Alan Marcus, Adrian Monck, Thomas Philbeck and Philip Shetler-Jones.

My deep gratitude also goes to all members of the Forum community who helped shape my thinking about the fourth industrial revolution. I am particularly thankful to Andrew McAfee and Erik Brynjolfsson for inspiring my ideas on the impact of technological innovation and the great challenges and opportunities that lie ahead, and to Dennis Snower and Stewart Wallis for underscoring the need for values-based narratives if we are to succeed in harnessing the fourth industrial revolution for the global good.

Additional thanks to Marc Benioff, Katrine Bosley, Justine Cassell, Mariette DiChristina, Murali Doraiswamy, Nita Farahany, Zev Furst, Nik Gowing, Victor Halberstadt, Ken Hu, Lee Sang-Yup, Alessio Lomuscio, Jack Ma, Ellen

MacArthur, Peter Maurer, Bernard Meyerson, Andrew Maynard, William McDonough, James Moody, Andrew Moore, Michael Osborne, Fiona Paua Schwab, Feike Sijbesma, Vishal Sikka, Philip Sinclair, Hilary Sutcliffe, Nina Tandon, Farida Vis, Sir Mark Walport and Alex Wyatt, all of whom I corresponded with or were interviewed for this book.

The Forum's Network of Global Agenda Councils and our "future-oriented communities" strongly engaged in this topic and provided rich insights on all the topics discussed here. Special appreciation goes to the Global Agenda Councils on the Future of Software and Society, Migration and the Future of Cities. I am also grateful to the remarkable array of thought leaders who so generously contributed their time and insights on this topic during the Summit on the Global Agenda 2015 in Abu Dhabi, as well as members of the Forum's Global Shapers, Young Global Leaders, and Young Scientists communities, particularly those who contributed ideas through TopLink, the Forum's virtual knowledge and collaboration platform.

Special thanks also to Alejandro Reyes for his editing, Scott David for the design, and Kamal Kimaoui for his layouts and publishing touch.

To have the book ready in time for the Annual Meeting 2016, it had to be written in less than three months with the collaboration of people all over the world. This truly reflects the fast-paced, dynamic environment of the fourth industrial revolution. So lastly, I convey my deep gratitude to you, the reader, for embarking on this journey with me, and for your enduring commitment to improving the state of the world.

Appendix: Deep Shift

In the fourth industrial revolution, digital connectivity enabled by software technologies is fundamentally changing society. The scale of the impact and the speed of the changes taking place have made the transformation that is playing out so different from any other industrial revolution in human history.

The World Economic Forum's Global Agenda Council on the Future of Software and Society conducted a survey of 800 executives to gauge when business leaders anticipate that these game-changing technologies would break into the public domain to a significant degree, and to understand fully the implications of these shifts to individuals, organizations, government and society.

The survey report *Deep Shift – Technology Tipping Points and Social Impact* was published in September 2015.[76] Reproduced below are 21 technology shifts presented in the study and two additional ones, including the tipping points for these technologies and the dates of their expected arrival to market.

Shift 1: Implantable Technologies

The tipping point: The first implantable mobile phone available commercially

By 2025: 82% of respondents expected this tipping point will have occurred

People are becoming more and more connected to devices, and those devices are increasingly becoming connected to their bodies. Devices are not just being worn, but also being implanted into bodies, serving communications, location and behaviour monitoring, and health functions.

Pacemakers and cochlear implants were just the beginning of this, with many more health devices constantly being launched. These devices will be able to sense the parameters of diseases; they will enable individuals to take action, send data to monitoring centres, or potentially release healing medicines automatically.

Smart tattoos and other unique chips could help with identification and location. Implanted devices will likely also help to communicate thoughts normally expressed verbally through a "built-in" smart phone, and potentially unexpressed thoughts or moods by reading brainwaves and other signals.

Positive impacts

– Reduction in missing children
– Increased positive health outcomes
– Increased self-sufficiency
– Better decision-making
– Image recognition and availability of personal data (anonymous network that will "yelp"[77] people)

Negative impacts

– Privacy/potential surveillance
– Decreased data security
– Escapism and addiction
– Increased distractions (i.e. attention deficit disorder)

Unknown, or cuts both ways

— Longer lives
— Changing nature of human relationships
— Changes in human interactions and relationships
— Real-time identification
— Cultural shift (eternal memory)

The shift in action

— Digital tattoos not only look cool but can perform useful tasks, like unlocking a car, entering mobile phone codes with a finger-point or tracking body processes.

Source: https://wtvox.com/3d-printing-in-wearable-tech/top-10-implantable-wearables-soon-body/

— According to a WT VOX article: "Smart Dust, arrays of full computers with antennas, each much smaller than a grain of sand, can now organize themselves inside the body into as-needed networks to power a whole range of complex internal processes. Imagine swarms of these attacking early cancer, bringing pain relief to a wound or even storing critical personal information in a manner that is deeply encrypted and hard to hack. With smart dust, doctors will be able to act inside your body without opening you up, and information could be stored inside you, deeply encrypted, until you unlock it from your very personal nano network."

Source: https://wtvox.com/3d-printing-in-wearable-tech/top-10-implantable-wearables-soon-body/

— A smart pill, developed by Proteus Biomedical and Novartis, has a biodegradable digital device attached to it, which transmits data to your phone on how the body is interacting with the medication.

Source: http://cen.acs.org/articles/90/i7/Odd-Couplings.html

Shift 2: Our Digital Presence

The tipping point: 80% of people with a digital presence on the internet

By 2025: 84% of respondents expected this tipping point will have occurred

Having a presence in the digital world has evolved rapidly in the past 20 or more years. Just 10 years ago, it meant having a mobile phone number, email address and perhaps a personal website or a MySpace page.

Now, people's digital presence is regarded as their digital interactions, and traces through a multitude of online platforms and media. Many people have more than one digital presence, such as a Facebook page, Twitter account, LinkedIn profile, Tumblr blog, Instagram account and often many more.

In our increasingly connected world, digital life is becoming inextricably linked with a person's physical life. In the future, building and managing a digital presence will become as common as when people decide how to present themselves to the world everyday through fashion, words and acts. In that connected world and through their digital presence, people will be able to seek and share information, freely express ideas, find and be found, and develop and maintain relationships virtually anywhere in the world.

Positive impacts

- Increased transparency
- Increased and faster interconnection between individuals and groups
- Increase in free speech
- Faster information dissemination/exchange
- More efficient use of government services

Negative impacts

- Privacy/potential surveillance
- More identity theft
- Online bullying/stalking
- Groupthink within interest groups and increased polarization
- Disseminating inaccurate information (the need for reputation management); echo chambers[78]
- Lack of transparency where individuals are not privy to information algorithms (for news/information)

Unknown, or cuts both ways

- Digital legacies/footprints
- More targeted advertising
- More targeted information and news
- Individual profiling
- Permanent identity (no anonymity)
- Ease of developing online social movement (political groups, interest groups, hobbies, terrorist groups)

The shift in action

If the three largest popular social media sites were countries, they would have almost a billion more people than China "See Figure I."

Figure I: Active Users of Social Media sites compared with the populations of the world's largest countries

Top 10 Populations ('000,000)

1		**Facebook**	**1,400**
2		China	1,360
3		India	1,240
4		**Twitter**	**646**
5		USA	318
6		Indonesia	247
7		Brazil	202
8		Pakistan	186
9		Nigeria	173
10		**Instagram**	**152**

Source: http://mccrindle.com.au/the-mccrindle-blog/social-media-and-narcissism

Shift 3: Vision as the New Interface

The tipping point: 10% of reading glasses connected to the internet

By 2025: 86% of respondents expected this tipping point will have occurred

Google Glass is just the first of many potential ways in which glasses, eyewear/headsets and eye-tracking devices can become "intelligent" and lead to eyes and vision being the connection to the internet and connected devices.

With direct access to internet applications and data through vision, an individual's experiences can be enhanced, mediated or completely augmented to provide different, immersive reality. Also, with emerging eye-tracking technologies, devices can feed information through visual interfaces, and eyes can be the source for interacting with and responding to the information.

Enabling vision as an immediate, direct interface – by providing instruction, visualization and interaction – can change the way that learning, navigation, instruction and feedback for producing goods and services, experiencing entertainment and enabling the disabled are helping people to engage more fully with the world.

Positive impacts

- Immediate information to the individual to make informed decisions for navigation and work/personal activities
- Improved capacity to perform tasks or produce goods and services with visual aids for manufacturing, healthcare/surgery and service delivery
- Ability for those with disabilities to manage their interactions and movement, and to experience the world – through speaking, typing and moving, and via immersive experiences

Negative impacts

–   Mental distraction causing accidents
–   Trauma from negative immersive experiences
–   Increased addiction and escapism

Unknown, or cuts both ways

–   A new segment created in the entertainment industry
–   Increased immediate information

The shift in action

Glasses are already on the market today (not just produced by Google) that can:

–   Allow you to freely manipulate a 3D object, enabling it to be moulded like clay
–   Provide all the extended live information you need when you see something, in the same way the brain functions
–   Prompt you with an overlay menu of the restaurant you pass by
–   Project picture or video on any piece of paper

Source: http://www.hongkiat.com/blog/augmented-reality-smart-glasses/

Shift 4: Wearable Internet

The tipping point: 10% of people wearing clothes connected to the internet

By 2025: 91% of respondents expected this tipping point will have occurred

Technology is becoming increasingly personal. Computers were first located in large rooms, then on desks and, following that, on people's laps. While technology can now be found in people's mobile phones in their pockets, it will soon be integrated directly into clothing and accessories.

Released in 2015, Apple Watch is connected to the internet and contains many of the same functional capabilities as a smart phone. Increasingly, clothing and other equipment worn by people will have embedded chips that connect the article and person wearing it to the internet.

Positive impacts

– More positive health outcomes leading to longer lives
– More self-sufficiency
– Self-managed healthcare
– Better decision-making
– Decrease in missing children
– Personalized clothes (tailoring, design)

Negative impacts

– Privacy/potential surveillance
– Escapism/addiction
– Data security

Unknown, or cuts both ways

- Real-time identification
- Change in personal interactions and relationships
- Image recognition and availability of personal data (anonymous network that will "yelp" you)

The shift in action

The research and advisory group, Gartner, estimates approximately 70 million smart watches and other bands will be sold by in 2015, with the total increasing to 514 million within five years.

Source: http://www.zdnet.com/article/wearables-internet-of-thingsmuscle-in-on-smartphone-spotlight-at-mwc/

Mimo Baby has created a fast-growing wearable baby monitor that reports a baby's breathing, body position, sleep activity, etc., to your iPad or smart phone. (This has caused some controversy over where to draw the line between helping, and creating a solution to a problem that doesn't exist. In this case, supporters say it helps the baby sleep better, while critics say sensors are not a replacement for parenting.

Source: http://mimobaby.com/; http://money.cnn.com/2015/04/16/smallbusiness/mimo-wearable-baby-monitor/

Ralph Lauren has developed a sports shirt that is designed to provide real-time workout data by measuring sweat output, heart rate, breathing intensity, etc.

Source: http://www.ralphlauren.com/product/index.jsp?productId=69917696&ab=rd_men_features_thepolotechshirt&cp=64796626.65333296

Shift 5: Ubiquitous Computing

The tipping point: 90% of the population with regular access to the internet

By 2025: 79% of respondents expected this tipping point will have occurred

Computing is becoming more accessible every day, and computing power has never been more available to individuals – be that via a computer with internet connection, a smart phone with 3G/4G or services in the cloud.

Today, 43% of the world's population is connected to the internet.[79] And, 1.2 billion smart phones were sold in 2014 alone.[80] In 2015, sales of tablets are estimated to take over sales of personal computers (PCs), while mobile phone sales (all combined) will outpace computers by six to one.[81] As the internet has been outgrowing every other media channel in speed of adoption, it is expected that, in only a few years, three-quarters of the world's population will have regular access to the web.

In the future, regular access to the internet and information will no longer be a benefit of developed economies, but a basic right just like clean water. Because wireless technologies require less infrastructure than many other utilities (electricity, roads and water), they will very likely become accessible much quicker than the others. Hence, anyone from any country will be able to access and interact with information from the opposite corner of the world. Content creation and dissemination will become easier than ever before.

Positive impacts

- More economic participation of disadvantaged populations located in remote or underdeveloped regions ("last mile")
- Access to education, healthcare and government services
- Presence

- Access to skills, greater employment, shift in types of jobs
- Expanded market size/e-commerce
- More information
- More civic participation
- Democratization/political shifts
- "Last mile": increased transparency and participation versus an increase in manipulation and echo chambers

Negative impacts

- Increased manipulation and echo chambers
- Political fragmentation
- Walled gardens (i.e. limited environments, for authenticated users only) do not allow full access in some regions/countries

The shift in action

To make the internet available to the next 4 billion users, two key challenges must be overcome: access must be available, and it must be affordable. The race to provide the rest of the world access to the web is underway. Already, over 85% of the world's population lives within a couple kilometres of a mobile phone tower that could deliver internet service.[82] Mobile operators around the world are expanding internet access rapidly. Facebook's Internet.org, a project with mobile network operators, has enabled access to free basic internet services for over a billion people in 17 countries in the last year.[83] And, many initiatives are under way to affordably connect even the most remote regions: Facebook's Internet.org is developing internet drones, Google's Project Loon is using balloons and SpaceX is investing in new low-cost satellite networks.

Shift 6: A Supercomputer in Your Pocket

The tipping point: 90% of the population using smart phones

By 2025: 81% of respondents expected this tipping point will have occurred

Already in 2012, the Google Inside Search team published that "it takes about the same amount of computing to answer one Google Search query as all the computing done – in flight and on the ground – for the entire Apollo programme!"[84] Moreover, current smart phones and tablets contain more computing power than many of the formerly known supercomputers, which used to fill an entire room.

Global smart phone subscribers are anticipated to total 3.5 billion by 2019; that will equate to 59% smart phone penetration by population, surpassing the 50% penetration of 2017 and underlining the significant growth from the 28% level in 2013.[85] In Kenya, Safaricom, the leading mobile service operator, reported that 67% of handset sales were smart phones in 2014, and the GSMA forecasts that Africa will have over half a billion smart phone users by 2020.[86]

The shift in devices has already occurred in many countries across different continents (with Asia leading the trend today), as more people are using their smart phones rather than traditional PCs. As technology is progressing to miniaturize devices, increase computing power and, especially, decrease the price of electronics, smart phone adoption will only accelerate.

According to Google, the countries in Figure II have a higher usage of smart phones than PCs.

Figure II: Countries with Higher Smart Phone Usage than PC (March 2015)

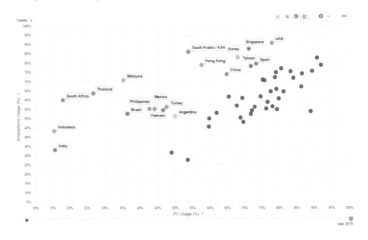

Source: http://www.google.com.sg/publicdata/explore

Figure III: Countries with Nearly 90% of Adult Population Using Smart Phones (March 2015)

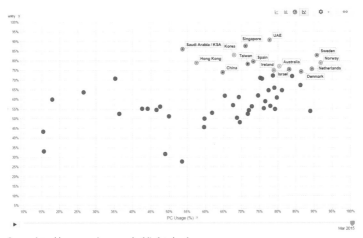

Source: http://www.google.com.sg/publicdata/explore

Countries such as Singapore, South Korea and the United Arab Emirates (UAE) are the closest to reaching the tipping point of 90% of the adult population using smart phones (Figure III).

Society is headed towards adopting even faster machines that will allow users to perform complicated tasks on the go. Most likely, the number of devices that each person uses will grow strongly, not only with new functions performed but also with specialization of tasks.

Positive Impacts

- More economic participation of disadvantaged populations located in remote or underdeveloped regions ("last mile")
- Access to education, healthcare and government services
- Presence
- Access to skills, greater employment, shift in types of jobs
- Expanded market size/e-commerce
- More information
- More civic participation
- Democratization/political shifts
- "Last mile": increased transparency and participation versus an increase in manipulation and echo chambers

Negative impacts

- Increased manipulation and echo chambers
- Political fragmentation
- Walled gardens (i.e. limited environments, for authenticated users only) do not allow full access in some regions/countries

Unknown, or cuts both ways

- 24/7 – always on
- Lack of division between business and personal
- Be anywhere/everywhere
- Environmental impact from manufacturing

The shift in action

In 1985, the Cray-2 supercomputer was the fastest machine in the world. The iPhone 4, released in June 2010, had the power equivalent to the Cray-2; now, the Apple Watch has the equivalent speed of two iPhone 4s just five years later.[87] With the consumer retail price of smart phones tumbling to below $50, processing power skyrocketing and adoption in emerging markets accelerating, nearly everyone will soon have a literal supercomputer in their pocket.

Source: http://pages.experts-exchange.com/processing-power-compared/

Shift 7: Storage for All

The tipping point: 90% of people having unlimited and free (advertising-supported) storage

By 2025: 91% of respondents expected this tipping point to have occurred

Storage capabilities have evolved tremendously in the past years, with an increasing number of companies offering it almost for free to their users as part of the service benefits. Users are producing increasing amounts of content, without worrying about ever having to delete it to make room for more. A clear trend of commoditizing storage capacity exists. One reason for it is that the storage price (Figure IV) has dropped exponentially (by a factor of approximately ten, every five years).

Figure IV: Hard Drive Cost per Gigabyte (1980-2009)

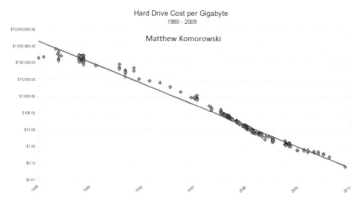

Source: "a history of storage costs", mkomo.com, 8 September 2009[88]

An estimated 90% of the world's data has been created in the past two years, and the amount of information created by businesses is doubling every 1.2 years.[89] Storage has already become a commodity, with companies like Amazon Web Services and Dropbox leading this trend.

The world is heading towards a full commoditization of storage, through free and unlimited access for users. The best-case scenario of revenue for companies could potentially be advertising or telemetry.

Positive impacts

- Legal systems
- History scholarship/academia
- Efficiency in business operations
- Extension of personal memory limitations

Negative impact

- Privacy surveillance

Unknown, or cuts both ways
- Eternal memory (nothing deleted)
- Increased content creation, sharing and consumption

The shift in action

Numerous companies already offer free storage in the cloud, ranging from 2 GB to 50 GB.

Shift 8: The Internet of and for Things

The tipping point: 1 trillion sensors connected to the internet

By 2025: 89% of respondents expected this tipping point to have occurred

With continuously increasing computing power and falling hardware prices (still in line with Moore's Law[90]), it is economically feasible to connect literally anything to the internet. Intelligent sensors are already available at very competitive prices. All things will be smart and connected to the internet, enabling greater communication and new data-driven services based on increased analytics capabilities.

A recent study looked into how sensors can be used to monitor animal health and behaviour.[91] It demonstrates how sensors wired in cattle can communicate to each other through a mobile phone network, and can provide real-time data on cattle conditions from anywhere.

Experts suggest that, in the future, every (physical) product could be connected to ubiquitous communication infrastructure, and sensors everywhere will allow people to fully perceive their environment.

Positive impacts

- Increased efficiency in using resources
- Rise in productivity
- Improved quality of life
- Effect on the environment
- Lower cost of delivering services
- More transparency around the use and state of resources
- Safety (e.g. planes, food)
- Efficiency (logistics)
- More demand for storage and bandwidth
- Shift in labour markets and skills
- Creation of new businesses

- Even hard, real-time applications feasible in standard communication networks
- Design of products to be "digitally connectable"
- Addition of digital services on top of products
- Digital twin provides precise data for monitoring, controlling and predicting
- Digital twin becomes active participant in business, information and social processes
- Things will be enabled to perceive their environment comprehensively, and react and act autonomously
- Generation of additional knowledge, and value based on connected "smart" things

Negative impacts

- Privacy
- Job losses for unskilled labour
- Hacking, security threat (e.g. utility grid)
- More complexity and loss of control

Unknown, or cuts both ways

- Shift in business model: asset rental/usage, not ownership (appliances as a service)
- Business model impacted by the value of the data
- Every company potentially a software company
- New businesses: selling data
- Change in frameworks to think about privacy
- Massively distributed infrastructure for information technologies
- Automation of knowledge work (e.g. analyses, assessments, diagnoses)
- Consequences of a potential "digital Pearl Harbor" (i.e. digital hackers or terrorists paralysing infrastructure, leading to no food, fuel and power for weeks)
- Higher utilization rates (e.g. cars, machines, tools, equipment, infrastructure)

The shift in action

The Ford GT has 10 million lines of computer code in it.

Source: http://rewrite.ca.com/us/articles/security/iot-is-bringing-lots-ofcode-to-your-car-hackers-too.html?intcmp=searchresultclick&resultnum=2).

The new model of the popular VW Golf has 54 computer processing units; as many as 700 data points get processed in the vehicle, generating six gigabytes of data per car.

Source: "IT-Enabled Products and Services and IoT", Roundtable on Digital Strategies Overview, Center for Digital Strategies at the Tuck School of Business at Dartmouth, 2014

More than 50 billion devices are expected to be connected to the internet by 2020. Even the Milky Way, the earth's galaxy, contains only around 200 billion suns!

Eaton Corporation builds sensors into certain high-pressure hoses that sense when the hose is about to fray, preventing potentially dangerous accidents and saving the high costs of downtime of the machines that have the hoses as a key component.

Source: "The Internet of Things: The Opportunities and Challenges of Interconnectedness", Roundtable on Digital Strategies Overview, Center for Digital Strategies at the Tuck School of Business at Dartmouth, 2014

Already last year, according to BMW 8% of cars worldwide, or 84 million, were connected to the internet in some way., That number will grow to 22%, or 290 million cars, by 2020.

Source: http://www.politico.eu/article/google-vs-german-car-engineerindustry-american-competition/

Insurance companies like Aetna are thinking about how sensors in a carpet could help if you've had a stroke. They would detect any gait change and have a physical therapist visit.

Source: "The Internet of Things: The Opportunities and Challenges of Interconnectedness", Roundtable on Digital Strategies Overview, Center for Digital Strategies at the Tuck School of Business at Dartmouth, 2014

Shift 9: The Connected Home

Tipping point: Over 50% of internet traffic delivered to homes for appliances and devices (not for entertainment or communication)

By 2025: 70% of respondents expected this tipping point to have occurred

In the 20th century, most of the energy going into a home was for direct personal consumption (lighting). But over time, the amount of energy used for this and other needs was eclipsed by much more complex devices, from toasters and dishwashers to televisions and air conditioners.

The internet is going the same way: most internet traffic to homes is currently for personal consumption, in communication or entertainment. Moreover, very fast changes are already occurring in home automation, enabling people to control lights, shades, ventilation, air conditioning, audio and video, security systems and home appliances. Additional support is provided by connected robots for all kinds of services – as, for example, vacuum cleaning.

Positive impacts

- Resource efficiency (lower energy use and cost)
- Comfort
- Safety/security, and detection of intrusions
- Access control
- Home sharing
- Ability to live independently (young/old, those disabled)
- Increased targeted advertising and overall impact on business
- Reduced costs of healthcare systems (fewer hospital stays and physician visits for patients, monitoring the drug-taking process)

- Monitoring (in real-time) and video recording
- Warning, alarming and emergency requests
- Remote home control (e.g. close the gas valve)

Negative impacts

- Privacy
- Surveillance
- Cyber attacks, crime, vulnerability

Unknown, or cuts both ways

- Impact on workforce
- Change in work's location (more from and outside the home)
- Privacy, data ownership

The shift in action

An example of this development for use in the home was cited by cnet.com:

"Nest, makers of the Internet-connected thermostat and smoke detector … announced [in 2014] the 'Works with Nest' developer program, which makes sure products from different companies work with its software. For example, a partnership with Mercedes Benz means your car can tell Nest to turn up the heat at home so it's warm when you arrive … Eventually … hubs like Nest's will help the home sense what you need, adjusting everything automatically. The devices themselves might eventually disappear into the home, merely acting as sensors and devices controlled from a single hub."

Source: "Rosie or Jarvis: The future of the smart home is still in the air", Richard Nieva, 14 January 2015, cnet.com, http://www.cnet.com/news/rosie-or-jarvisthe-future-of-the-smart-home-is-still-in-the-air/

Shift 10: Smart Cities

Tipping point: The first city with more than 50,000 inhabitants and no traffic lights

By 2025: 64% of respondents expected this tipping point to have occurred

Many cities will connect services, utilities and roads to the internet. These smart cities will manage their energy, material flows, logistics and traffic. Progressive cities, such as Singapore and Barcelona, are already implementing many new data-driven services, including intelligent parking solutions, smart trash collection and intelligent lighting. Smart cities are continuously extending their network of sensor technology and working on their data platforms, which will be the core for connecting the different technology projects and adding future services based on data analytics and predictive modelling.

Positive impacts

- Increased efficiency in using resources
- Rise in productivity
- Increased density
- Improved quality of life
- Effect on the environment
- Increased access to resources for the general population
- Lower cost of delivering services
- More transparency around the use and state of resources
- Decreased crime
- Increased mobility
- Decentralized, climate friendly energy production and consumption
- Decentralized production of goods
- Increased resilience (to impacts of climate change)
- Reduced pollution (air, noise)
- Increased access to education

- Quicker/speed up accessibility to markets
- More employment
- Smarter e-government

Negative impacts

- Surveillance, privacy
- Risk of collapse (total black out) if the energy system fails
- Increased vulnerability to cyber attacks

Unknown, or cuts both ways

- Impact on city culture and feel
- Change of individual habitus of cities

The shift in action

According to a paper published in The Future Internet:

"The city of Santander in northern Spain has 20,000 sensors connecting buildings, infrastructure, transport, networks and utilities. The city offers a physical space for experimentation and validation of functions, such as interaction and management protocols, device technologies, and support services such as discovery, identity management and security".

Source: "Smart Cities and the Future Internet: Towards Cooperation Frameworks for Open Innovation", H. Schaffers, N. Komninos, M. Pallot, B. Trousse, M. Nilsson and A. Oliveira, The Future Internet, J. Domingue et al. (eds), LNCS 6656, 2011, pp. 431-446, http://link.springer.com/chapter/10.1007%2F978-3-642-20898-0_31

Shift 11: Big Data for Decisions

The tipping point: The first government to replace its census with big-data sources

By 2025: 83% of respondents expected this tipping point to have occurred

More data exists about communities than ever before. And, the ability to understand and manage this data is improving all the time. Governments may start to find that their previous ways of collecting data are no longer needed, and may turn to big-data technologies to automate their current programmes and deliver new and innovative ways to service citizens and customers.

Leveraging big data will enable better and faster decision-making in a wide range of industries and applications. Automated decision-making can reduce complexities for citizens and enable businesses and governments to provide real-time services and support for everything from customer interactions to automated tax filings and payments.

The risks and opportunities in leveraging big data for decision-making are significant. Establishing trust in the data and algorithms used to make decisions will be vital. Citizen concerns over privacy and establishing accountability in business and legal structures will require adjustments in thinking, as well as clear guidelines for use in preventing profiling and unanticipated consequences. Leveraging big data to replace processes that today are done manually may render certain jobs obsolete, but may also create new categories of jobs and opportunities that currently do not exist in the market.

Positive impacts

- Better and faster decisions
- More real-time decision-making
- Open data for innovation
- Jobs for lawyers
- Reduced complexity and more efficiency for citizens
- Cost savings
- New job categories

Negative impacts

- Job losses
- Privacy concerns
- Accountability (who owns the algorithm?)
- Trust (how to trust data?)
- Battles over algorithms

Unknown, or cuts both ways

- Profiling
- Change in regulatory, business and legal structures

The shift in action

The volume of business data worldwide, across all companies, doubles every 1.2 years.

Source: "A Comprehensive List of Big Data Statistics," Vincent Granville, 21 October 2014: http://www.bigdatanews.com/profiles/blogs/acomprehensive-list-of-big-data-statistics

"Farmers from Iowa to India are using data from seeds, satellites, sensors, and tractors to make better decisions about what to grow, when to plant, how to track food freshness from farm to fork, and how to adapt to changing climates."

Source: "What's the Big Deal with Data", BSA | Software Alliance, http://data.bsa.org/

"To better inform restaurant-goers about unsanitary venues, San Francisco successfully piloted a collaboration with Yelp—fusing the city's restaurant health inspection data onto the site's restaurant review pages. If you open up the page of restaurant Tacos El Primo, for example, it shows a health score of 98 out of 100 (below). Yelp ratings are pretty powerful. Apart from serving as a mouthpiece for the city to tell residents about food hazards, the collaboration is potentially a way to shame repeat-offender restaurants into complying with health standards."

Source: http://www.citylab.com/cityfixer/2015/04/3-cities-using-opendata-in-creative-ways-to-solve-problems/391035/

Shift 12: Driverless Cars

The tipping point: Driverless cars equalling 10% of all cars on US roads

By 2025: 79% of respondents expected this tipping point to have occurred

Trials of driverless cars from large companies such as Audi and Google are already taking place, with a number of other enterprises ramping up efforts to develop new solutions. These vehicles can potentially be more efficient and safer than cars with people behind the steering wheel. Moreover, they could reduce congestion and emissions, and upend existing models of transportation and logistics.

Positive impacts

– Improved safety
– More time for focusing on work and/or consuming media content
– Effect on the environment
– Less stress and road rage
– Improved mobility for those older and disabled, among others
– Adoption of electric vehicles

Negative impacts

– Job losses (taxi and truck drivers, car industry)
– Upending of insurance and roadside assistance ("pay more to drive yourself")
– Decreased revenue from traffic infringements
– Less car ownership
– Legal structures for driving
– Lobbying against automation (people not allowed to drive on freeways)
– Hacking/cyber attacks

The shift in action

In October 2015, Tesla made its cars that were sold over the last year in the US semi-autonomous via a software update.

Source: http://www.wired.com/2015/10/tesla-self-driving-over-air-update-live

Google plans to make autonomous cars available to the public in 2020.

Source: Thomas Halleck, 14 January 2015, "Google Inc. Says Self-Driving Car Will Be Ready By 2020", International Business Times: http://www.ibtimes.com/google-inc-says-self-driving-car-will-be-ready-2020-1784150

In the summer of 2015, two hackers demonstrated their ability to hack into a moving car, controlling its dashboard functions, steering, brakes etc., all through the vehicle's entertainment system.

Source: http://www.wired.com/2015/07/hackers-remotely-kill-jeep-highway/

The first state in the United States (Nevada) to pass a law allowing driverless (autonomous) cares did so in 2012.

Source: Alex Knapp, 22 June 2011, "Nevada Passes Law Authorizing Driverless Cars", Forbes: http://www.forbes.com/sites/alexknapp/2011/06/22/nevadapasses-law-authorizing-driverless-cars/

Shift 13: Artificial Intelligence and Decision-Making

The tipping point: The first Artificial Intelligence (AI) machine on a corporate board of directors

By 2025: 45% of respondents expected this tipping point to have occurred

Beyond driving cars, AI can learn from previous situations to provide input and automate complex future decision processes, making it easier and faster to arrive at concrete conclusions based on data and past experiences.

Positive impacts

– Rational, data-driven decisions; less bias
– Removal of "irrational exuberance"
– Reorganization of outdated bureaucracies
– Job gains and innovation
– Energy independence
– Advances in medical science, disease eradication

Negative impacts

– Accountability (who is responsible, fiduciary rights, legal)
– Job losses
– Hacking/cybercrime
– Liability and accountability, governance
– Becoming incomprehensible
– Increased inequality
– "Falling foul of the algorithm"
– Existential threat to humanity

The shift in action

ConceptNet 4, a language AI, recently passed an IQ test better than most four-year-olds – three years ago it could barely compete with a one-year-old. The next version, just finalized, is expected to perform on level with a five- to six year-old.

Source: "Verbal IQ of a Four-Year Old Achieved by an AI System": http://citeseerx.ist.psu.edu/viewdoc/download?doi=10.1.1.386.6705&rep=rep1&type=pdf

If Moore's Law continues to develop with the same speed as it has been for the past 30 years, CPUs will reach the same level of processing power as the human brain in 2025. Deep Knowledge Ventures, a Hong Kong-based venture capital fund that invests in life sciences, cancer research, age-related diseases and regenerative medicine has appointed an artificial intelligence algorithm called VITAL (Validating Investment Tool for Advancing Life Sciences) to its board of directors.

Source: "Algorithm appointed board director", BBC: http://www.bbc.com/news/technology-27426942

Shift 14: AI and White-Collar Jobs

The tipping point: 30% of corporate audits performed by AI

By 2025: 75% of respondents expected this tipping point to have occurred

AI is good at matching patterns and automating processes, which makes the technology amenable to many functions in large organizations. An environment can be envisioned in the future where AI replaces a range of functions performed today by people.

An Oxford Martin School study[92] looked into the susceptibility of jobs to computerization from AI and robotics, and came up with some sobering results. Their model predicted that up to 47% of US jobs in 2010 were highly likely to become computerized in the next 10-20 years (Figure V).

Figure V: Distribution of US Occupational Employment* over the Probability of Computerization

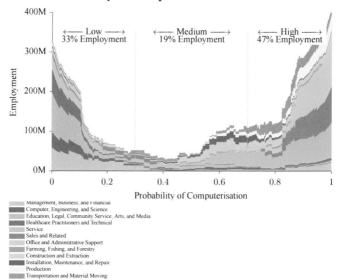

* Distribution based on 2010 job mix.
Source: Frey, C.B. and M.A. Osborne, "The Future of Employment: How Susceptible Are Jobs to Computerisation?", 17 September 2013

Positive impacts

– Cost reductions
– Efficiency gains
– Unlocking innovation, opportunities for small business, start-ups (smaller barriers to entry, "software as a service" for everything)

Negative impacts

– Job losses
– Accountability and liability
– Change to legal, financial disclosure, risk
– Job automation (refer to the Oxford Martin study)

The shift in action

Advances in automation were reported on by FORTUNE:

"IBM's Watson, well known for its stellar performance in the TV game show Jeopardy!, has already demonstrated a far more accurate diagnosis rate for lung cancers than humans – 90% versus 50% in some tests. The reason is data. Keeping pace with the release of medical data could take doctors 160 hours a week, so doctors cannot possibly review the amount of new insights or even bodies of clinical evidence that can give an edge in making a diagnosis. Surgeons already use automated systems to aid in low-invasive procedures."

In Erik Sherman, FORTUNE, 25 February 2015, http://fortune.com/2015/02/25/5-jobs-that-robots-already-are-taking/

Shift 15: Robotics and Services

The tipping point: The first robotic pharmacist in the US

By 2025: 86% of respondents expected this tipping point to have occurred

Robotics is beginning to influence many jobs, from manufacturing to agriculture, and retail to services. According to the International Federation of Robotics, the world now includes 1.1 million working robots, and machines account for 80% of the work in manufacturing a car.[93] Robots are streamlining supply chains to deliver more efficient and predictable business results.

Positive impacts

– Supply chain and logistics, eliminations
– More leisure time
– Improved health outcomes (big data for pharmaceutical gains in research and development)
– Banking ATM as early adopter
– More access to materials
– Production "re-shoring" (i.e. replacing overseas workers with robots)

Negative impacts

– Job losses
– Liability, accountability
– Day-to-day social norms, end of 9-to-5 and 24-hour services
– Hacking and cyber-risk

The shift in action

An article from The Fiscal Times appearing on CNBC.com states that:

"Rethink Robotics released Baxter [in the fall of 2012] and received an overwhelming response from the manufacturing industry, selling out of their production capacity through April ...

[In April] Rethink launch[ed] a software platform that [allows] Baxter to do a more complex sequencing of tasks – for example, picking up a part, holding it in front of an inspection station and receiving a signal to place it in a 'good' or 'not good' pile. The company also [released] a software development kit ... that will allow third parties – like university robotics researchers – to create applications for Baxter."

In "The Robot Reality: Service Jobs Are Next to Go", Blaire Briody, 26 March 2013, The Fiscal Times, http://www.cnbc.com/id/100592545

Shift 16: Bitcoin and the Blockchain

The tipping point: 10% of global gross domestic product (GDP) stored on blockchain technology

By 2025: 58% of respondents expected this tipping point to have occurred

Bitcoin and digital currencies are based on the idea of a distributed trust mechanism called the "blockchain", a way of keeping track of trusted transactions in a distributed fashion. Currently, the total worth of bitcoin in the blockchain is around $20 billion, or about 0.025% of global GDP of around $80 trillion.

Positive impacts

– Increased financial inclusion in emerging markets, as financial services on the blockchain gain critical mass
– Disintermediation of financial institutions, as new services and value exchanges are created directly on the blockchain
– An explosion in tradable assets, as all kinds of value exchange can be hosted on the blockchain
– Better property records in emerging markets, and the ability to make everything a tradable asset
– Contacts and legal services increasingly tied to code linked to the blockchain, to be used as unbreakable escrow or programmatically designed smart contracts
– Increased transparency, as the blockchain is essentially a global ledger storing all transactions

The shift in action

Smartcontracts.com provides programmable contracts that do payouts between two parties once certain criteria have been met, without involving a middleman. These contracts are secured in the blockchain as "self-executing contractual states", which eliminate the risk of relying on others to follow through on their commitments.

Shift 17: The Sharing Economy

The tipping point: Globally more trips/journeys via car sharing than in private cars

By 2025: 67% of respondents expected this tipping point to have occurred

The common understanding of this phenomenon is the usually technology-enabled ability for entities (individuals or organizations) to share the use of a physical good/asset, or share/provide a service, at a level that was not nearly as efficient or perhaps even possible before. This sharing of goods or services is commonly possible through online marketplaces, mobile apps/location services or other technology-enabled platforms. These have reduced the transaction costs and friction in the system to a point where it is an economic gain for all involved, divided in much finer increments.

Well-known examples of the sharing economy exist in the transportation sector. Zipcar provides one method for people to share use of a vehicle for shorter periods of time and more reasonably than traditional rental car companies. RelayRides provides a platform to locate and borrow someone's personal vehicle for a period of time. Uber and Lyft provide much more efficient "taxi-like" services from individuals, but aggregated through a service, enabled by location services and accessed through mobile apps. In addition, they are available at a moment's notice.

The sharing economy has any number of ingredients, characteristics or descriptors: technology enabled, preference for access over ownership, peer to peer, sharing of personal assets (versus corporate assets), ease of access, increased social interaction, collaborative consumption and openly shared user feedback (resulting in increased trust). Not all are present in every "sharing economy" transaction.

Positive impacts

- Increased access to tools and other useful physical resources
- Better environmental outcomes (less production and fewer assets required)
- More personal services available
- Increased ability to live off cash flow (with less need for savings to be able to afford use of assets)
- Better asset utilization
- Less opportunity for long-term abuse of trust because of direct and public feedback loops
- Creation of secondary economies (Uber drivers delivering goods or food)

Negative impacts

- Less resilience after a job loss (because of less savings)
- More contract / task-based labour (versus typically more stable long-term employment)
- Decreased ability to measure this potentially grey economy
- More opportunity for short-term abuse of trust
- Less investment capital available in the system

Unknown, or cuts both ways

- Changed property and asset ownership
- More subscription models
- Less savings
- Lack of clarity on what "wealth" and "well off" mean
- Less clarity on what constitutes a "job"
- Difficulty in measuring this potentially "grey" economy
- Taxation and regulation adjusting from ownership/sales-based models to use-based models

The shift in action

A particular notion of ownership underlies this development and is reflected in the following questions.

- The largest retailer doesn't own a single store? (Amazon)
- The largest provider of sleeping rooms doesn't own a single hotel? (Airbnb)
- The largest provider of transportation doesn't own a single car? (Uber)

Shift 18: Governments and the Blockchain

The tipping point: Tax collected for the first time by a government via a blockchain

By 2025: 73% of respondents expected this tipping point to have occurred

The blockchain creates both opportunities and challenges for countries. On the one hand, it is unregulated and not overseen by any central bank, meaning less control over monetary policy. On the other hand, it creates the ability for new taxing mechanisms to be built into the blockchain itself (e.g. a small transaction tax).

Unknown impacts, or cut both ways

– Central banks and monetary policy
– Corruption
– Real-time taxation
– Role of government

The shift in action

In 2015, the first virtual nation, BitNation, was created using blockchain as the foundation identification technology for citizen's identity cards. At the same time, Estonia became the first real government to deploy the blockchain technology.

Sources: https://bitnation.co/; http://www.pymnts.com/news/2014/estoniannational-id-cards-embrace-electronic-payment-capabilities/#.Vi9T564rJPM

Shift 19: 3D Printing and Manufacturing

The tipping point: The first 3D-printed car in production

By 2025: 84% of respondents expected this tipping point to have occurred

3D printing, or additive manufacturing, is the process of creating a physical object by printing it layer upon layer from a digital 3D drawing or model. Imagine creating a loaf of bread, slice by slice. 3D printing has the potential to create very complex products without complex equipment.[94] Eventually, many different kinds of materials will be used in the 3D printer, such as plastic, aluminium, stainless steel, ceramic or even advanced alloys, and the printer will be able to do what a whole factory was once required to accomplish. It is already being used in a range of applications, from making wind turbines to toys.

Over time, 3D printers will overcome the obstacles of speed, cost and size, and become more pervasive. Gartner has developed a "Hype Cycle" chart (Figure VI) showing the various stages of different 3D printing capabilities and their market impact, and plotting most business uses of the technology as entering the "slope of enlightenment".[95]

Figure VI: Hype Cycle for 3D Printing

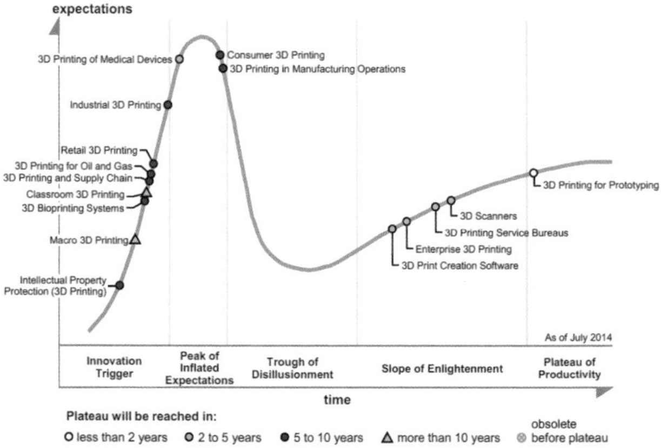

Source: Gartner (July 2014)

Positive impacts

- Accelerated product development
- Reduction in the design-to-manufacturing cycle
- Easily manufactured intricate parts (not possible or difficult to do earlier)
- Rising demand for product designers
- Educational institutions using 3D printing to accelerate learning and understanding
- Democratized power of creation/manufacturing (both limited only by the design)
- Traditional mass manufacturing responding to the challenge by finding ways to reduce costs and the size of minimum runs
- Growth in open-source "plans" to print a range of objects
- Birth of a new industry supplying printing materials
- Rise in entrepreneurial opportunities in the space[96]
- Environmental benefits from reduced transportation requirements

Negative impacts

– Growth in waste for disposal, and further burden on the environment
– Production of parts in the layer process that are anisotropic, i.e. their strength is not the same in all directions, which could limit the functionality of parts
– Job losses in a disrupted industry
– Primacy of intellectual property as a source of value in productivity
– Piracy
– Brand and product quality

Unknown, or cuts both ways

– Potential that any innovation can be instantly copied

The shift in action

An example of 3D printing for manufacturing has been recently covered by FORTUNE:

"General Electric's Leap jet engine is not only one of the company's bestsellers, it's going to incorporate a fuel nozzle produced entirely through additive manufacturing. The process, popularly known as 3-D printing, involves building up layers of material (in this case alloyed metals) according to precise digital plans. GE is currently completing testing of the new Leap engines, but the benefit of additive manufactured parts has already been proven on other models."

Source: "GE's first 3D-printed parts take flight", Andrew Zaleski, FORTUNE, 12 May 2015, http://fortune.com/2015/05/12/ge-3d-printed-jet-engine-parts/

Shift 20: 3D Printing and Human Health

The tipping point: The first transplant of a 3D-printed liver

By 2025: 76% of respondents expected this tipping point to have occurred

One day, 3D printers may create not only things, but also human organs – a process called bioprinting. In much the same process as for printed objects, an organ is printed layer by layer from a digital 3D model.[97] The material used to print an organ would obviously be different from what is used to print a bike, and experimenting can be done with the kinds of materials that will work, such as titanium powder for making bones. 3D printing has great potential to service custom design needs; and, there is nothing more custom than a human body.

Positive impacts

- Addressing the shortage of donated organs (an average of 21 people die each day waiting for transplants that can't take place because of the lack of an organ)[98]
- Prosthetic printing: limb/body part replacements
- Hospitals printing for each patient requiring surgery (e.g. splints, casts, implants, screws)
- Personalized medicine: 3D printing growing fastest where each customer needs a slightly different version of a body part (e.g. a crown for a tooth)
- Printing components of medical equipment that are difficult or expensive to source, such as transducers[99]
- Printing, for example, dental implants, pacemakers and pens for bone fracture at local hospitals instead of importing them, to reduce the cost of operations

- Fundamental changes in drug testing, which can be done on real human objects given the availability of fully printed organs
- Printing of food, thus improving food security

Negative impacts

- Uncontrolled or unregulated production of body parts, medical equipment or food

- Growth in waste for disposal, and further burden on the environment
- Major ethical debates stemming from the printing of body parts and bodies: Who will control the ability to produce them? Who will ensure the quality of the resulting organs?
- Perverted disincentives for health: If everything can be replaced, why live in a healthy way?
- Impact on agriculture from printing food

The shift in action

The first use of a 3D-printed spine implant was reported by Popular Science:

"[In 2014], doctors at Peking University Third Hospital successfully implanted the first ever 3-D-printed section of vertebra into [a] young patient to replace a cancerous vertebra in his neck. The replacement vertebra was modelled from the boy's existing vertebra, which made it easier for them to integrate.

Source: "Boy Given a 3-D Printed Spine Implant, Loren Grush, Popular Science, 26 August 2014, http://www.popsci.com/article/science/boy-given-3-dprinted-spine-implant

Shift 21: 3D Printing and Consumer Products

The tipping point: 5% of consumer products printed in 3D

By 2025: 81% of respondents expected this tipping point to have occurred

Because 3D printing can be done by anyone with a 3D printer, it creates opportunities for typical consumer products to be printed locally and on demand, instead of having to be bought at shops. A 3D printer will eventually be an office or even a home appliance. This further reduces the cost of accessing consumer goods and increases the availability of 3D printed objects. Current usage areas for 3D printing (Figure VII) indicate several sectors related to developing and producing consumer products (proof of concept, prototype and production).

Figure VII: Use of 3D Printing in Various Areas (% of respondents*)

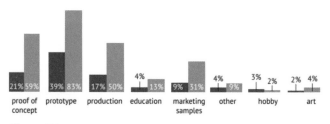

| | proof of concept | prototype | production | education | marketing samples | other | hobby | art |

■ Total ▨ Power users

* Percentages are of respondents from the Sculpteo survey.

Source: Sculpteo, The State of 3D Printing (survey of 1,000 people), as published in Hedstrom, J., "The State of 3D Printing…", Quora[100]

Positive impacts

- More personalized products and personal fabrication
- Creating niche products, and making money selling them
- Fastest growth of 3D printing where each customer has slightly different needs from a product – e.g. a particular shaped foot requires a specially sized shoe
- Reduced logistics costs, with the possibility of huge energy savings[101]
- Contributing to abundant local activities; crafting own goods that benefit from the removal of logistics costs (circular economy)

Negative impacts

- Global and regional supply and logistics chain: lower demand resulting in job losses
- Gun control: opening opportunities for printing objects with high levels of abuse, such as guns
- Growth in waste for disposal, and further burden on the environment
- Major disruption of production controls, consumer regulations, trade barriers, patents, taxes and other government restrictions; and, the struggle to adapt

The shift in action

Almost 133,000 3D printers were shipped worldwide in 2014, a 68% increase from 2013. The majority of printers, selling for under $10,000, are thus suitable for applications from labouratories and schools to small manufacturing businesses. As a result, the size of the 3D materials and services industry grew strongly, to $3.3 billion.[102]

Shift 22: Designer Beings[103]

The tipping point: The first human whose genome was directly and deliberately edited is born

Since the turn of the century, the cost of sequencing an entire human genome has fallen by almost six orders of magnitude. The human genome project spent $2.7 billion to produce the first entire genome in 2003. By 2009 the cost per genome was down to 100k while today it is possible for researchers to pay a lab specialising in such matters only $1000 to sequence a human genome. A similar trend has occurred more recently in genome editing with the development of the CRISPR/Cas9 method, which is being widely adopted due to its higher effectiveness and efficiency and lower cost than previous approaches.

The real revolution is hence not the sudden ability for dedicated scientists to edit the genes of plants and animals, but rather the increased ease that new sequencing and editing technologies provide, vastly increasing the number of researchers who are able to conduct experiments

Positive impacts

- Higher agricultural yields thanks to crops and crop treatments which are more robust, effective and productive
- More effective medical therapies via personalised medicine
- Faster, more accurate, less invasive medical diagnostics
- Higher levels of understanding of human impact on nature
- Reduced incidence of genetic disease and related suffering

Negative impacts

– Risk of interaction between edited plants/animals human/ environmental health
– Exacerbated inequality due to high cost of access to therapies
– Social backlash or rejection of gene editing technologies
– Misuse of genetic data by governments or companies
– International disagreements about ethical use of genome editing technologies

Unknown or cuts both ways

– Increased longevity
– Ethical dilemmas regarding nature of humanity
– Cultural shifts

The shift in action

 "In March 2015, leading scientists publish a Nature article calling for a moratorium on editing human embryos, highlighting "grave concerns regarding the ethical and safety implications of this research". Only one month later, in April 2015, "Researchers led by Junjiu Huang of Yat-sen University in Guangzhou published the world's first scientific paper on altering the DNA of human embryos."

Sources: http://www.nature.com/news/don-t-edit-the-human-germ-line-1.17111; http://qz.com/389494/chinese-researchers-are-the-first-to-genetically-modify-a-human-embryo-and-many-scientists-think-theyve-gone-too-far/http://qz.com/389494/chinese-researchers-are-the-first-to-genetically-modify-a-human-embryo-and-many-scientists-think-theyve-gone-too-far/

Shift 23: Neurotechnologies[104]

The tipping point: The first human with fully artificial memory implanted in the brain

There is not one area of our personal and professional lives that cannot benefit from a better understanding of how our brain functions – at both the individual and collective levels. This is underscored by the fact that – over the past few years - two of the most funded research programs in the world are in brain sciences: The *Human Brain Project* (a €1 billion project over 10 years funded by the European Commission) and President Obama's *Brain Research Through Advancing Innovative Neurotechnologies* (BRAIN) Initiative. Although these programs are primarily focused on scientific and medical research, we are also witnessing the rapid growth (and influence) of neurotechnologies in non-medical aspects of our lives. Neurotechnology consists of monitoring brain activity and looking at how the brain changes and/or interfaces with the world.

In 2015, for example, the portability and the affordability of neuro-headsets (which already cost less than a gaming console) offer unprecedented possibilities - marking what is likely to be not only a neuro-revolution, but also a societal one[105].

Positive impacts

– Disabled people can now control prosthetic limbs or wheel-chairs "with their minds".
– Neurofeedback, the possibility to monitor brain activity in real time, offers countless possibilities to help fight addictions, regulate food behaviour, and improve performances ranging from sports to the classroom.
– Being able to collect, process, store and compare large amounts of brain activity-related data allows us to improve diagnosis and treatment efficiency of brain disorders and mental health-related issues.

- The law will be able to provide customized processing on cases and address responsibility issues in criminal cases in a differential fashion rather than in a generic one now.
- The next generation of computers, whose design has been informed by brain science, may reason, predict and react just like the human cortex (an area of the brain known as the seat of intelligence).

Negative impacts

- Brain-based discrimination: Individuals are not just their brains, as such there is a risk for decisions to be made in a context-independent fashion, based only on brain data in fields ranging from the law to HR, consumer behaviour or education[106].
- Fear of what thoughts/dreams/desires to be decrypted and for privacy to no longer exist,
- Fear of creativity or the human touch to slowly but surely disappear, mainly carried so far by overselling what brain sciences can do.
- Blurring the lines between man and machine

Unknown, or cuts both ways

- Cultural shift
- Disembodiment of communication
- Improvement of performance
- Extending human cognitive abilities will trigger new behaviours

The shift in action

- Cortical computing algorithms have already shown an ability to solve modern CAPTCHAs (widely used tests to distinguish humans from machines).
- The automotive industry has developed systems monitoring attention and awareness that can stop cars when people are falling asleep while driving.

- An intelligent computer program in China scored better than many human adults on an IQ test.
- IBM's Watson supercomputer, after sifting through millions of medical records and databases, has begun to help doctors choose treatment options for patients with complex needs.
- Neuromorphic image sensors, i.e. inspired how the eye and brain communicate, will have impact ranging from battery usage to robotics
- Neuroprosthetics are allowing disabled people to control artificial members and exoskeletons. Some blind people will be able to see (again).
- The Restoring Active Memory (RAM) program by DARPA is a precursor to memory restoration and enhancement
- Depression symptoms in mice could be cured by the artificial reactivation of happy memories as evidenced by Neuroscientists at MIT

Doraiswamy M. (2015). 5 brain technologies that will shape our future. World Economic Forum Agenda, Aug 9
https://agenda.weforum.org/2015/08/5-brain-technologies-future/

Fernandez A (2015). 10 neurotechnologies about to transform brain enhancement and brain health. SharpBrains, USA, Nov 10
http://sharpbrains.com/blog/2015/11/10/10-neurotechnologies-about-to-transform-brain-enhancement-and-brain-health/

Notes

[1] The terms "disruption" and "disruptive innovation" have been much discussed in business and management strategy circles, most recently in Clayton M. Christensen, Michael E. Raynor, and Rory McDonald, *What is Disruptive Innovation?,* Harvard Business Review, December 2015. While respecting the concerns of Professor Christensen and his colleagues about definitions, I have employed the broader meanings in this book.

[2] Erik Brynjolfsson and Andrew McAfee, *The Second Machine Age: Work, Progress, and Prosperity in a Time of Brilliant Technologies,* W.W. Norton & Company, 2014.

[3] James Manyika and Michael Chui, "Digital Era Brings Hyperscale Challenges", *The Financial Times,* 13 August 2014.

[4] The designer and architect Neri Oxman offers a fascinating example of what I just described. Her research lab works at the intersection of computational design, additive manufacturing, materials engineering and synthetic biology.
https://www.ted.com/talks/neri_oxman_design_at_the_intersection_of_technology_and_biology

[5] Carl Benedikt Frey and Michael Osborne, with contributions from Citi Research, "Technology at Work – The Future of Innovation and Employment", Oxford Martin School and Citi, February 2015. https://ir.citi.com/jowGiIw%2FoLrkDA%2BldI1U%2FYUEpWP9ifowg%2F4HmeO9kYfZ-iN3SeZwWEvPez7gYEZXmxsFM7eq1gc0%3D

[6] David Isaiah, "Automotive grade graphene: the clock is ticking", *Automotive World,* 26 August 2015.
http://www.automotiveworld.com/analysis/automotive-grade-graphene-clock-ticking/

[7] Sarah Laskow, "The Strongest, Most Expensive Material on Earth", *The Atlantic,*
http://www.theatlantic.com/technology/archive/2014/09/the-strongest-most-expensive-material-on-earth/380601/

[8] Some of the technologies are described in greater detail in: Bernard Meyerson, "Top 10 Technologies of 2015", Meta-Council on Emerging Technologies, World Economic Forum, 4 March 2015.
https://agenda.weforum.org/2015/03/top-10-emerging-technologies-of-2015-2/

[9] Tom Goodwin, "In the age of disintermediation the battle is all for the consumer interface", *TechCrunch*, March 2015.
http://techcrunch.com/2015/03/03/in-the-age-of-disintermediation-the-battle-is-all-for-the-customer-interface/

[10] K.A. Wetterstrand, "DNA Sequencing Costs: Data from the NHGRI Genome Sequencing Program (GSP)", National Human Genome Research Institute, 2 October 2015.
http://www.genome.gov/sequencingcosts/

[11] Ariana Eunjung Cha, "Watson's Next Feat? Taking on Cancer", *The Washington Post,* 27 June 2015.
http://www.washingtonpost.com/sf/national/2015/06/27/watsons-next-feat-taking-on-cancer/

[12] Jacob G. Foster, Andrey Rzhetsky and James A. Evans, "Tradition and Innovation in Scientists' Research Strategies", *American Sociological Review*, October 2015 80: 875-908
http://www.knowledgelab.org/docs/1302.6906.pdf

[13] Mike Ramsay and Douglas Cacmillan, "Carnegie Mellon Reels After Uber Lures Away Researchers", *Wall Street Journal,* 31 May 2015
http://www.wsj.com/articles/is-uber-a-friend-or-foe-of-carnegie-mellon-in-robotics-1433084582

[14] World Economic Forum, *Deep Shift – Technology Tipping Points and Societal Impact*, Survey Report, Global Agenda Council on the Future of Software and Society, September 2015.

[15] For more details on the survey methodology, please refer to pages 4 and 39 of the report referenced in the previous note.

[16] UK Office of National Statistics, "Surviving to Age 100", 11 December 2013, http://www.ons.gov.uk/ons/rel/lifetables/historic-and-projected-data-from-the-period-and-cohort-life-tables/2012-based/info-surviving-to-age-100.html

[17] The Conference Board, *Productivity Brief 2015*, 2015.
According to data compiled by The Conference Board data, global labour productivity growth in the period 1996-2006 averaged 2.6%, compared to 2.1% for both 2013 and 2014.
https://www.conference-board.org/retrievefile.cfm?filename=The-Conference-Board-2015-Productivity-Brief.pdf&type=subsite

[18] United States Department of Labor, "Productivity change in the nonfarm business sector, 1947-2014", Bureau of Labor Statistics
http://www.bls.gov/lpc/prodybar.htm

[19] United States Department of Labor, "Preliminary multifactor productivity trends, 2014", Bureau of Labor Statistics, 23 June 2015
http://www.bls.gov/news.release/prod3.nr0.htm

[20] OECD, "The Future of Productivity", July 2015. http://www.oecd.org/eco/growth/The-future-of-productivity-policy-note-July-2015.pdf
For a short discussion on decelerating US productivity, see: John Fernald and Bing Wang, "The Recent Rise and Fall of Rapid Productivity Growth", Federal Reserve Bank of San Francisco, 9 February 2015.
http://www.frbsf.org/economic-research/publications/economic-letter/2015/february/economic-growth-information-technology-factor-productivity/

[21] The economist Brad DeLong makes this point in: J. Bradford DeLong, "Making Do With More", Project Syndicate, 26 February 2015.
http://www.project-syndicate.org/commentary/abundance-without-living-standards-growth-by-j--bradford-delong-2015-02

[22] John Maynard Keynes, "Economic Possibilities for our Grandchildren" in Essays in Persuasion, Harcourt Brace, 1931.

[23] Carl Benedikt Frey and Michael Osborne, "The Future of Employment: How Susceptible Are Jobs to Computerisation?", Oxford Martin School, Programme on the Impacts of Future Technology, University of Oxford, 17 September 2013. http://www.oxfordmartin.ox.ac.uk/downloads/academic/The_Future_of_Employment.pdf

[24] Shelley Podolny, "If an Algorithm Wrote This, How Would You Even Know?", The New York Times, 7 March 2015
http://www.nytimes.com/2015/03/08/opinion/sunday/if-an-algorithm-wrote-this-how-would-you-even-know.html?_r=0

[25] Martin Ford, *Rise of the Robots,* Basic Books, 2015.

[26] Daniel Pink, *Free Agent Nation – The Future of Working for Yourself,* Grand Central Publishing, 2001.

[27] Quoted in: Farhad Manjoo, "Uber's business model could change your work", *The New York Times,* 28 January 2015.

[28] Quoted in: Sarah O'Connor, "The human cloud: A new world of work", *The Financial Times,* 8 October 2015.

[29] Lynda Gratton, *The Shift: The Future of Work is Already Here,* Collins, 2011.

[30] R. Buckminster Fuller and E.J. Applewhite, *Synergetics: Explorations in the Geometry of Thinking,* Macmillan, 1975.

[31] Eric Knight, "The Art of Corporate Endurance", Harvard Business Review, April 2, 2014
https://hbr.org/2014/04/the-art-of-corporate-endurance

[32] VentureBeat, "WhatsApp now has 700M users, sending 30B messages per day", January 6 2015
http://venturebeat.com/2015/01/06/whatsapp-now-has-700m-users-sending-30b-messages-per-day/

[33] Mitek and Zogby Analytics, *Millennial Study 2014* , September 2014
https://www.miteksystems.com/sites/default/files/Documents/zogby_final_embargo_14_9_25.pdf

[34] Gillian Wong, "Alibaba Tops Singles' Day Sales Record Despite Slowing China Economy", The Wall Street Journal, 11 November 2015, http://www.wsj.com/articles/alibaba-smashes-singles-day-sales-record-1447234536

[35] "The Mobile Economy: Sub-Saharan Africa 2014", GSM Association, 2014.
http://www.gsmamobileeconomyafrica.com/GSMA_ME_SubSaharanAfrica_Web_Singles.pdf

[36] Tencent, "Announcement of results for the three and nine months ended 30 September 2015"
http://www.tencent.com/en-us/content/ir/an/2015/attachments/20151110.pdf

[37] MIT, "The ups and downs of dynamic pricing", innovation@work Blog, MIT Sloan Executive Education, 31 October 2014.
http://executive.mit.edu/blog/the-ups-and-downs-of-dynamic-pricing#.VG4yA_nF-bU

[38] Giles Turner, "Cybersecurity Index Beat S&P500 by 120%. Here's Why, in Charts", Money Beat, *The Wall Street Journal,* 9 September 2015.
http://blogs.wsj.com/moneybeat/2015/09/09/cybersecurity-index-beats-sp-500-by-120-heres-why-in-charts/

[39] IBM, "Redefining Boundaries: Insights from the Global C-Suite Study," November 2015.
http://www-935.ibm.com/services/c-suite/study/

[40] Global e-Sustainability Initiative and The Boston Consulting Group, Inc, "GeSI SMARTer 2020: The Role of ICT in Driving a Sustainable Future", December 2012.
http://gesi.org/SMARTer2020

[41] Moisés Naím, *The End of Power: From Boardrooms to Battlefields and Churches to States, Why Being in Charge Isn't What It Used to Be,* Basic Books, 2013.
The book attributes the end of power to three revolutions: the "more" revolution, the mobility revolution, and the mentality revolution. It is careful in not identifying the role of information technology as predominant but there is no doubt that the more, the mobility and the mentality owe a lot to the digital age and the diffusion of new technologies.

[42] This point is made and developed in: "The Middle Kingdom Galapagos Island Syndrome: The Cul-De-Sac of Chinese Technology Standards", Information Technology and Innovation Foundation (ITIF), 15 December 2014.
http://www.itif.org/publications/2014/12/15/middle-kingdom-galapagos-island-syndrome-cul-de-sac-chinese-technology

[43] "Innovation Union Scoreboard 2015", European Commission, 2015. http://ec.europa.eu/growth/industry/innovation/facts-figures/scoreboards/files/ius-2015_en.pdf The measurement framework used in the Innovation Union Scoreboard distinguishes between three main types of indicators and eight innovation dimensions, capturing a total of 25 different indicators. The enablers capture the main drivers of innovation performance external to the firm and cover three innovation dimensions: human resources; open, excellent and attractive research systems; and finance and support. Firm activities capture the innovation efforts at the level of the firm, grouped in three innovation dimensions: firm investments, linkages and entrepreneurship, and intellectual assets. Outputs cover the effects of firms' innovation activities in two innovation dimensions: innovators and economic effects.

[44] World Economic Forum, *Collaborative Innovation – Transforming Business, Driving Growth*, August 2015. http://www3.weforum.org/docs/WEF_Collaborative_Innovation_report_2015.pdf

[45] World Economic Forum, *Global Information Technology Report 2015: ICTs for Inclusive Growth*, Soumitra Dutta, Thierry Geiger and Bruno Lanvin, eds., 2015.

[46] World Economic Forum, *Data-Driven Development: Pathways for Progress*, January 2015 http://www3.weforum.org/docs/WEFUSA_DataDrivenDevelopment_Report2015.pdf

[47] Tom Saunders and Peter Baeck, "Rethinking Smart Cities From The Ground Up", Nesta, June 2015. https://www.nesta.org.uk/sites/default/files/rethinking_smart_cities_from_the_ground_up_2015.pdf

[48] Carolina Moreno, "Medellin, Colombia Named 'Innovative City Of The Year' In WSJ And Citi Global Competition", Huffington Post, 2 March 2013 http://www.huffingtonpost.com/2013/03/02/medellin-named-innovative-city-of-the-year_n_2794425.html

[49] World Economic Forum, *Top Ten Urban Innovations*, Global Agenda Council on the Future of Cities, World Economic Forum, October 2015. http://www3.weforum.org/docs/Top_10_Emerging_Urban_Innovations_report_2010_20.10.pdf

[50] Alex Leveringhaus and Gilles Giacca, "Robo-Wars – The Regulation of Robotic Weapons", The Oxford Institute for Ethics, Law and Armed Conflict, The Oxford Martin Programme on Human Rights for Future Generations, and The Oxford Martin School, 2014. http://www. oxfordmartin.ox.ac.uk/downloads/briefings/Robo-Wars.pdf

[51] James Giordano quoted in Tom Requarth, "This is Your Brain. This is Your Brain as a Weapon", *Foreign Policy,* 14 September 2015. http://foreignpolicy.com/2015/09/14/this-is-your-brain-this-is-your-brain-as-a-weapon-darpa-dual-use-neuroscience/

[52] Manuel Castells, "The impact of the Internet on Society: A Global Perspective", *MIT Technology Review,* 8 September 2014. http://www.technologyreview.com/view/530566/the-impact-of-the-internet-on-society-a-global-perspective/

[53] Credit Suisse, *Global Wealth Report 2015,* October 2015. http://publications.credit-suisse.com/tasks/render/file/index. cfm?fileid=F2425415-DCA7-80B8-EAD989AF9341D47E

[54] OECD, "Divided We Stand: Why Inequality Keeps Rising", 2011. http://www.oecd.org/els/soc/49499779.pdf

[55] Frederick Solt, "The Standardized World Income Inequality Database," Working paper, SWIID, Version 5.0, October 2014. http://myweb.uiowa.edu/fsolt/swiid/swiid.html

[56] Richard Wilkinson and Kate Pickett, *The Spirit Level: Why Greater Equality Makes Societies Stronger*, Bloomsbury Press, 2009.

[57] Sean F. Reardon and Kendra Bischoff, "More unequal and more separate: Growth in the residential segregation of families by income, 1970-2009", US 2010 Project, 2011. http://www.s4.brown.edu/us2010/Projects/Reports.htm http://cepa.stanford.edu/content/more-unequal-and-more-separate-growth-residential-segregation-families-income-1970-2009

[58] Eleanor Goldberg, "Facebook, Google are Saving Refugees and Migrants from Traffickers", *Huffington Post,* 10 September 2015. http://www.huffingtonpost.com/entry/facebook-google-maps-refugeesmigrants_55f1aca8e4b03784e2783ea4

[59] Robert M. Bond, Christopher J. Fariss, Jason J. Jones, Adam D. I. Kramer, Cameron Marlow, Jaime E. Settle, and James H. Fowler, "A 61-million-person experiment in social influence and political mobilization", *Nature,* 2 September 2012 (online).
http://www.nature.com/nature/journal/v489/n7415/full/nature11421.html

[60] Stephen Hawking, Stuart Russell, Max Tegmark, Frank Wilczek, "Stephen Hawking: 'Transcendence looks at the implications of artificial intelligence – but are we taking AI seriously enough?", *The Independent,* 2 May 2014.
http://www.independent.co.uk/news/science/stephen-hawking-transcendence-looks-at-the-implications-of-artificial-intelligence-but-are-we-taking-9313474.html

[61] Greg Brockman, Ilya Sutskever & the OpenAI team, "Introducing OpenAI", 11 December 2015
https://openai.com/blog/introducing-openai/

[62] Steven Levy, "How Elon Musk and Y Combinator Plan to Stop Computers From Taking Over", 11 December 2015
https://medium.com/backchannel/how-elon-musk-and-y-combinator-plan-to-stop-computers-from-taking-over-17e0e27dd02a#.qjj55npcj

[63] Sara Konrath, Edward O'Brien, and Courtney Hsing. "Changes in dispositional empathy in American college students over time: A meta-analysis." *Personality and Social Psychology Review* (2010).

[64] Quoted in: Simon Kuper, "Log out, switch off, join in", *FT Magazine,* 2 October 2015. http://www.ft.com/intl/cms/s/0/fc76fce2-67b3-11e5-97d0-1456a776a4f5.html

[65] Sherry Turkle, *Reclaiming Conversation: The Power of Talk in a Digital Age,* Penguin, 2015.

[66] Nicholas Carr, *The Shallows: How the Internet is changing the way we think, read and remember,* Atlantic Books, 2010.

[67] Pico Iyer, *The Art of Stillness: Adventures in Going Nowhere,* Simon and Schuster, 2014.

[68] Quoted in: Elizabeth Segran, "The Ethical Quandaries You Should Think About the Next Time You Look at Your Phone", *Fast Company,* 5 October 2015.
http://www.fastcompany.com/3051786/most-creative-people/the-ethical-quandaries-you-should-think-about-the-next-time-you-look-at

[69] The term "contextual intelligence" was coined by Nihtin Nohria several years before he became the dean of Harvard Business School.

[70] Klaus Schwab, *Moderne Unternehmensführung im Maschinenbau (Modern Enterprise Management in Mechanical Engineering)*, VDMA, 1971.

[71] Quoted in: Peter Snow, *The Human Psyche in Love, War & Enlightenment,* Boolarong Press, 2010.

[72] Daniel Goleman, "What Makes A Leader?", *Harvard Business Review,* January 2004.
https://hbr.org/2004/01/what-makes-a-leader

[73] Rainer Maria Rilke, *Letters to a Young Poet*, Insel Verlag, 1929.

[74] Voltaire wrote in French: *"Le doute n'est pas une condition agréable, mais la certitude est absurde."* "On the Soul and God", letter to Frederick William, Prince of Prussia, 28 November 1770, in S.G. Tallentyre, trans., *Voltaire in His Letters: Being a Selection from His Correspondence*, G.P. Putnam's Sons, 1919.

[75] Martin Nowak with Roger Highfield, *Super Cooperators: Altruism, Evolution, and Why We Need Each Other to Succeed,* Free Press, 2012.

[76] World Economic Forum, *Deep Shift – Technology Tipping Points and Societal Impact*, Survey Report, Global Agenda Council on the Future of Software and Society, November 2015

[77] Borrowing from the concept of the yelp.com website, in that people would be able to provide reviews directly to others and those reviews would be recorded and/or shared online through chips implanted in them.

[78] "Echo chamber" connotes those who unquestioningly agree with another person or who repeat what other people have said without thinking or questioning.

[79] Internet live stats, "Internet users in the world",
http://www.internetlivestats.com/internet-users/
http://www.worldometers.info/world-population/

[80] "Gartner Says Worldwide Traditional PC, Tablet, Ultramobile and Mobile Phone Shipments to Grow 4.2 Percent in 2014", Gartner, 7 July 2014.
http://www.gartner.com/newsroom/id/2791017

[81] "Number of smartphones sold to end users worldwide from 2007 to 2014 (in million units)", statista, 2015.
http://www.statista.com/statistics/263437/globalsmartphone-sales-to-end-users-since-2007/

[82] Lev Grossman, "Inside Facebook's Plan to Wire the World," Time, 15 December 2014.
http://time.com/facebook-world-plan/

[83] "One Year In: Internet.org Free Basic Services," Facebook Newsroom, 26 July 2015. http://newsroom.fb.com/news/2015/07/one-year-in-internet-org-free-basic-services/

[84] Udi Manber and Peter Norvig, "The power of the Apollo missions in a single Google search", Google Inside Search, 28 August 2012.
http://insidesearch.blogspot.com/2012/08/the-power-of-apollo-missions-in-single.html

[85] Satish Meena, "Forrester Research World Mobile And Smartphone Adoption Forecast, 2014 To 2019 (Global)," Forrester Research, 8 August 2014. https://www.forrester.com/Forrester+Research+World+Mobile+And+Smartphone+Adoption+Forecast+2014+To+2019+Global/fulltext/-/E-RES118252

[86] GSMA, "New GSMA Report Forecasts Half a Billion Mobile Subscribers in Sub-Saharan Africa by 2020", 6 November 2014.
http://www.gsma.com/newsroom/press-release/gsma-report-forecasts-half-a-billion-mobile-subscribers-ssa-2020/

[87] "Processing Power Compared: Visualizing a 1 trillion-fold increase in computing performance", Experts Exchange.
http://pages.experts-exchange.com/processing-power-compared/

[88] "A history of storage costs", mkomo.com, 8 September 2009
http://www.mkomo.com/cost-per-gigabyte
According to the website, data was retrieved from Historical Notes about the Cost of Hard Drive Storage Space (http://ns1758.ca/winch/winchest.html). Data from 2004 to 2009 was retrieved using Internet Archive Wayback Machine (http://archive.org/web/web.php).

[89] Elana Rot, "How Much Data Will You Have in 3 Years?", Sisense, 29 July 2015. http://www.sisense.com/blog/much-data-will-3-years/

[90] Moore's Law generally states that processor speeds, or the overall number of transistors in a central processing unit, will double every two years.

[91] Kevin Mayer, Keith Ellis and Ken Taylor, "Cattle Health Monitoring Using Wireless Sensor Networks", Proceedings of the Communication and Computer Networks Conference, Cambridge, MA, USA, 2004. http://www.academia.edu/781755/Cattle_health_monitoring_using_ wireless_sensor_networks

[92] Carl Benedikt Frey and Michael A. Osborne, "The Future of Employment: How Susceptible Are Jobs to Computerisation?", 17 September 2013. http://www.oxfordmartin.ox.ac.uk/downloads/academic/The_Future_of_ Employment.pdf

[93] Will Knight, "This Robot Could Transform Manufacturing," MIT Technology Review, 18 September 2012. http://www.technologyreview.com/news/429248/this-robotcould- transform-manufacturing/

[94] See http://www.stratasys.com/.

[95] Dan Worth, "Business use of 3D printing is years ahead of consumer uptake", V3.co.uk, 19 August 2014. http://www.v3.co.uk/v3-uk/news/2361036/business-use-of-3d-printing-is- years-ahead-of-consumer-uptake

[96] "The 3D Printing Startup Ecosystem", SlideShare.net, 31 July 2014. http://de.slideshare.net/SpontaneousOrder/3d-printing-startup-ecosystem

[97] Alban Leandri, "A Look at Metal 3D Printing and the Medical Implants Industry", 3DPrint.com, 20 March 2015. http://3dprint.com/52354/3d-print-medical-implants/

[98] "The Need is Real: Data", US Department of Health and Human Services, organdonor.gov. http://www.organdonor.gov/about/data.html

[99] "An image of the future", The Economist, 19 May 2011. http://www.economist.com/node/18710080

[100] Jessica Hedstrom, "The State of 3D Printing", 23 May 2015. http:// jesshedstrom.quora.com/The-State-of-3D-Printing

[101] Maurizio Bellemo, "The Third Industrial Revolution: From Bits Back to Atoms", CrazyMBA.Club, 25 January 2015.
http://www.crazymba.club/the-third-industrial-revolution/

[102]T.E. Halterman, "3D Printing Market Tops $3.3 Billion, Expands by 34% in 2014", 3DPrint.com, 2 April 2015.
http://3dprint.com/55422/3d-printing-market-tops-3-3-billion-expands-by-34-in-2014/

[103] Note: this tipping point was not a part of the original survey (Deep Shift – Technology Tipping Points and Societal Impact, Survey Report, World Economic Forum, September 2015)

[104] Ibid.

[105] Fernandez A, Sriraman N, Gurewitz B, Oullier O (2015). Pervasive neurotechnology: A groundbreaking analysis of 10,000+ patent filings transforming medicine, health, entertainment and business. SharpBrains, USA (206 p.)
http://sharpbrains.com/pervasive-neurotechnology/

[106] Oullier O (2012). Clear up this fuzzy thinking on brain scans. Nature, 483(7387), p. 7, doi: 10.1038/483007a
http://www.nature.com/news/clear-up-this-fuzzy-thinking-on-brain-scans-1.10127